# Smart Sensors, Measurement and Instrumentation

Volume 38

**Series Editor**

Subhas Chandra Mukhopadhyay, School of Engineering, Macquarie University, Sydney, NSW, Australia

More information about this series at http://www.springer.com/series/10617

Olfa Kanoun · Nabil Derbel
Editors

# Advanced Sensors for Biomedical Applications

Springer

*Editors*
Olfa Kanoun
Fakultät für Elektrotechnik und Informationstechnik
Technische Universität Chemnitz
Chemnitz, Germany

Nabil Derbel
Department of Electrical Engineering
Ecole Nationale d'Ingénieurs de Sfax
University of Sfax
Sfax, Tunisia

ISSN 2194-8402 ISSN 2194-8410 (electronic)
Smart Sensors, Measurement and Instrumentation
ISBN 978-3-030-71224-2 ISBN 978-3-030-71225-9 (eBook)
https://doi.org/10.1007/978-3-030-71225-9



This Springer imprint is published by the registered company Springer Nature Switzerland AG
The registered company address is: Gewerbestrasse 11, 6330 Cham, Switzerland

# Preface

In this book, we present recent developments in the field of biomedical sensors and systems covering a wide range of sensors, methods, systems, and instrumentation techniques for diagnosis, monitoring, treatment, and assistance. It provides insights into theory, applications, and perspectives relevant to the field of biomedical engineering, as well as the general paradigms and methodologies behind them. Each chapter provides enrichment of understanding on a research topic along with a balanced treatment of the relevant theories, methods, or applications. It reports on the latest advances achieved in the corresponding field of biomedical engineering. This book is a good reference for graduate students, researchers, educators, engineers, and scientists.

This book is dedicated to recent developments in the field of sensors and includes 10 chapters in total, structured into two parts as follows.

The first part of this book focuses on human activity analysis, and comprises five chapters:

- The first chapter entitled "A Survey of Human Action Recognition using Accelerometer Data" provides a detailed state-ofthe-art survey of the human action recognition by exploiting acceleration data.
- The second chapter entitled "Ultra Thin Nanocomposite In-Sole Pressure Sensor Matrix for Gait Analysis" focuses on the design and implementation of thin multi-walled carbon nanotubes (CNT)/polydimethylsil-oxane (PDMS) based nanocomposite pressure sensors for the analysis of the foot pressure distribution.
- The third chapter entitled "Piezo-Resistive Pressure and Strain Sensors for Biomedical and Tele-Manipulation Applications" is dedicated to sensors based on carbon materials, which are gaining importance due to their high flexibility, sensitivity, and medical compatibility. Specially, nanocomposite based pressure and strain sensors offer a potential tool that helps to build the basis for body-attached sensor networks.
- The fourth chapter entitled "Wireless Body Sensor Networks with Enhanced Reliability by Data Aggregation Based on Machine Learning Algorithms" presents a method for reducing energy consumption in a wireless body sensor network

by data aggregation, which reduced the necessity of re-transmission of data and improves at the same time the network lifetime.

- The fifth chapter entitled "Accelerated Moving Humans Detection Algorithm using Combined Global Descriptors on GPU Based on CUDA" proposes an implementation of human movement detection algorithm on GPU based on the programming language CUDA, in order to extract the image features using the Fourier descriptor on GPU, and based on the histogram of orient gradient descriptor on GPU.
- The sixth chapter entitled "Human Breathing Monitoring by Graphene Oxide Based Sensors" introduces an impedimetric breath sensor based on a graphene oxide film, deposited on a silver interdigitated electrode and a flexible substrate having a very fast and extremely high sensitivity to the relative humidity in the ambient.

The second part addresses electrochemical sensors for early screening and diagnosis of diseases as an emerging field with a special importance nowadays, and comprises four chapters:

- The seventh chapter entitled "Impedimetrics Detection of Human Interleukin 10 on Diazonium Salt Electroaddressed Gold Microelectrode Surfaces" reports on the development and the fabrication of gold microelectrodes based on silicon by silicon technology, for multiplex detection of cytokines, which became an important biomarker for the identification of end-stage heart failure during the early phase of left ventricular assisted device implantation.
- The eighth chapter entitled "Review on Recent Advances in Urinary Biomarkers Based Electrochemical Sensors for Prostate Cancer Detection" provides a review about recent advances in the design of electrochemical biosensors for the quantification of urinary biomarkers of prostate cancer, which is today one of the most frequently diagnosed malignancy in man.
- The ninth chapter entitled "Recent Advances in Ultrasensitive miRNA Biomarkers Detection" describes how MicroRNAs, giving insight into etiology and progression of several pathologies, can be detected by various biosensing technologies that have been developed recently with signal amplification using functional nanomaterials, nucleic acid circuitry, and enzymes.
- The tenth chapter entitled "Early Detection of Helicobacter Pylori Bacteria in Complex Samples" provides a review about several detection methods, such as stool antigen tests, fluorescent detection methods, calorimetric detection methods, surface plasmon resonance detection methods, as well as electrochemical methods.

Chemnitz, Germany — Olfa Kanoun
Sfax, Tunisia — Nabil Derbel
January 2021

# About This Book

The book highlights recent developments in the field of biomedical sensors with a focus on technology and design aspects of novel sensors and sensor systems.

Diagnosis plays a central role in healthcare and requires a variety of novel biomedical sensors and sensor systems. This creates an enormous ongoing demand for sensors for both everyday life as well as for medical care. Technologies concerning the analysis of human activities as well as for the early detection of diseases are moving into the focus of interest and form the basis for supporting human health and quality of life.

As such, the book offers a key reference guide about novel medical sensors and systems for students, engineers, sensors designers, and technicians.

# Contents

**A Survey of Human Action Recognition using Accelerometer Data** ..... 1
Amira Mimouna and Anouar Ben Khalifa
1 Introduction ..... 2
2 Accelerometer's Review and Applications ..... 4
3 HAR Using Accelerometer Data ..... 7
4 Datasets ..... 16
5 Fusion Framework ..... 20
6 Experimental Results and Analysis ..... 23
7 Conclusions ..... 29
References ..... 29

**Ultra Thin Nanocomposite In-Sole Pressure Sensor Matrix for Gait Analysis** ..... 33
Dhivakar Rajendran, Bilel Ben Atitallah, Rajarajan Ramalingame, Roberto Bautista Quijano Jose, and Olfa Kanoun
1 Introduction ..... 34
2 Gait Analysis ..... 34
3 CNT/Polymer Pressure Sensor ..... 36
4 Implementation and Evaluation ..... 39
5 Conclusion ..... 43
References ..... 44

**Piezo-Resistive Pressure and Strain Sensors for Biomedical and Tele-Manipulation Applications** ..... 47
Bilel Ben Atitallah, Dhivakar Rajendran, Zheng Hu, Rajarajan Ramalingame, Roberto Bautista Quijano Jose, Renato da Veiga Torres, Dhouha Bouchaala, Nabil Derbel, and Olfa Kanoun
1 Introduction ..... 48
2 Nanocomposite Sensors ..... 49
3 Hybrid Glove with Implemented Pressure and Strain Sensors ..... 56
4 Conclusion ..... 63
References ..... 63

**Wireless Body Sensor Networks with Enhanced Reliability by Data Aggregation Based on Machine Learning Algorithms** ................. 67
Mbarka Belhaj Mohamed, Amel Meddeb-Makhlouf, Ahmed Fakhfakh, and Olfa Kanoun
1 Introduction ........................................ 68
2 Related Works ........................................ 69
3 Proposed Model ........................................ 71
4 Performance Evaluation ........................................ 73
5 Discussion ........................................ 80
6 Conclusion ........................................ 80
References ........................................ 80

**Accelerated Moving Humans Detection Algorithm using Combined Global Descriptors on GPU Based on CUDA** ............... 83
Haythem Bahri, Marwa Chouchene, Randa Khemiri, Fatma Ezahra Sayadi, and Mohamed Atri
1 Introduction ........................................ 84
2 Methods and Algorithms ........................................ 85
3 Moving Humans Detection Algorithm and Its Implementation on GPU ........................................ 90
4 Experimental Results and Discussion ........................................ 92
5 Conclusion ........................................ 95
References ........................................ 96

**Human Breathing Monitoring by Graphene Oxide Based Sensors** ...... 97
Ammar Al-Hamry, Enza Panzardi, Marco Mugnaini, and Olfa Kanoun
1 Introduction ........................................ 98
2 Experimental Investigations ........................................ 99
3 Results and Discussion ........................................ 100
4 Conclusion ........................................ 106
References ........................................ 106

**Impedimetric Detection of Human Interleukin 10 on Diazonium Salt Electroaddressed Gold Microelectrode Surfaces** .................. 109
Michael Lee, Abdoullatif Baraket, Monique Sigaud, Ammar Al-Hamry, Nadia Zine, Olfa Kanoun, Joan Bausells, and Abdelhamid Errachid
1 Introduction ........................................ 110
2 Experimental Investigations ........................................ 112
3 Results and Discussion ........................................ 115
4 Conclusions ........................................ 119
References ........................................ 120

**Review on Recent Advances in Urinary Biomarkers Based Electrochemical Sensors for Prostate Cancer Detection** ............... 123
Meriem Mokni, Najla Fourati, Chouki Zerrouki, Ali Othmane, Asma Omezzine, and Ali Bouslama
1 Introduction ........................................................ 124
2 Prostate Cancer Antigen: PCA3 ........................................ 127
3 Sarcosine ........................................................... 128
4 Engrailed-2 Protein ................................................. 129
5 Micro-RNA (miRNA) ................................................... 130
6 $\alpha$-Methylacyl-CoA Racemase (AMACR) ............................ 131
7 Microseminoprotein-Beta (MSMB) ...................................... 131
8 TMPRSS2: ERG Transcripts ............................................ 132
9 Conclusion and Perspectives ......................................... 132
References ............................................................ 133

**Recent Advances in Ultrasensitive miRNA Biomarkers Detection** ...... 137
Khouloud Djebbi, Mohamed Bahri, Mohamed Amin Elaguech, Rong Tian, Shi Biao, Chaker Tlili, and Deqiang Wang
1 MicroRNAs Overview .................................................. 138
2 MicroRNA Detection .................................................. 141
References ............................................................ 159

**Early Detection of Helicobacter Pylori Bacteria in Complex Samples** ............................................................... 165
Hussamaldeen Jaradat, Ammar Al-Hamry, Mohammed Ibbini, and Olfa Kanoun
1 Introduction ........................................................ 166
2 Importance of HP In-Vitro Detection ................................. 167
3 Recent Advances in In-Vitro HP Detection ............................ 168
4 Conclusions and Future Perspective .................................. 173
References ............................................................ 174

# A Survey of Human Action Recognition using Accelerometer Data

**Amira Mimouna and Anouar Ben Khalifa**

**Abstract** Recognizing human actions and analyzing human behaviors from accelerometer data has become a challenging task. Hence, Human Action Recognition (HAR) using inertial sensors have been addressed in a plethora of various review papers. This chapter provides a detailed state of art survey of HAR exploiting acceleration data. Considering different modalities, we prove that the accelerometer is one of the most promising sensors in this field by presenting an overview of its applications. In addition, we propose a comprehensive review of recent studies in this domain along different views: from data modalities to feature extraction and classification steps. Moreover, we list the most publicly available databases that include accelerometer data. Afterwards, we used a multi-level fusion framework that includes signal-level, feature-level, score level and the decision level fusion in order to improve the recognition performance. For the classification, we took advantage of the support vector machine with features from the time-frequency domain. The proposed framework was evaluated using three public datasets: WARD, MHAD and Realdisp. The results obtained from the fusion techniques indicate that the score level provides a satisfactory performance compared to the other levels and with the use of each accelerometer separately.

**Keywords** Human action recognition · Accelerometer · Accelerometer applications · Multi-level fusion

A. Mimouna (✉) · A. Ben Khalifa
LATIS- Laboratory of Advanced Technology and Intelligent Systems, Université de Sousse, Ecole Nationale d'Ingénieurs de Sousse, Sousse, Tunisia
e-mail: amira.mimouna@yahoo.com

A. Ben Khalifa
e-mail: anouar.benkhalifa@eniso.rnu.tn

 
O. Kanoun and N. Derbel (eds.), *Advanced Sensors for Biomedical Applications*, Smart Sensors, Measurement and Instrumentation 38,
https://doi.org/10.1007/978-3-030-71225-9_1

# 1 Introduction

Do the human actions developed in a day present a good image of the overall homeostasis of the person? The execution of an action such as sitting down or standing for a long period, rather than jumping or lying down, and the speed of the accomplishment of these tasks present valuable information about a person's daily activity. It reflects his vitality, therefore his state of health and even his psychological state. Hence, monitoring and supervising the activities of everyday living has become a crucial task to enhance the quality of our lives.

Human actions can be classified into four classes, relying on their complexity: gestures, actions, interactions and a group of activities (Aggarwal and Ryoo 2011; Jegham et al. 2019). Gestures consist of elementary movements of a body part, for example: 'raising an arm'. Actions include gestures that are temporally ordered, for instance: 'walking' or 'waving'. In addition, interactions involve two persons or more as 'two persons shaking hands', also it exists a human-object interaction between humans and objects, such as 'a person is giving a cup to another'; finally, group of activities that include several persons and/or objects such as 'a group having a meeting'.

The principal aim of human action recognition is automatically detecting and analyzing human activities, then interpreting continuously and successfully the situation (Chen and Shen 2017). Thus, this filed of research has been unavoidable in several areas including: health care (Ameur et al. 2016; Jain and Kanhangad 2018), surveillance (Lejmi et al. 2017), human-computer interaction (Nuno et al. 2017), virtual reality (Kwon et al. 2017), gaming (Namal et al. 2006), etc.

To guarantee the recognition and the analysis of human behavior, several researchers have exploited different types of technologies in their work, including cameras, Kinect, accelerometers, gyroscopes, microphones, MoCap (motion capture), RFID (radio frequency identification), etc.

In fact, the employment of microphones in the field of human behavior analysis is becoming more and more important in various fields, such as robotic assistance, action recognition, etc. However; the presence of noise and the distance of the person with regard to the microphone are still a challenge (Rodomagoulakis et al. 2016).

Although several works have used RGB cameras because it provides rich information of the scene, the recognition based on video sequence has its own limitations, such as sensitivity to lighting, background disorder and occlusion (Jegham and Ben Khalifa 2017; Chebli and Ben Khalifa 2018). In addition, this approach is limited to a fixed area of view outlined by the camera position and for many people, who feel uncomfortable when they are monitored continuously, cameras are intrusive (Cornacchia et al. 2017; Lejmi et al. 2019).

Based on depth sensors that provide 3D action data, the human action recognition has gained more improvement. For the Kinect, it is insensitive to changes in lighting and ensures recognition of actions in the dark. Nevertheless, the subject must always be present in the field of view of the Kinect and the images present different noise.

**Table 1** A summary of some limitations of different sensors for human action recognition

| Sensors | Limitations |
|---|---|
| Microphone | – Presence of noise |
| | – The distance of the person with regard to the microphone |
| Camera | – Conditions of acquisition |
| | – Occlusions |
| | – Intrusive |
| Kinect | – The person must always be in the Kinect's point of view |
| | – The images present different noise |
| | – Requires different material sources |
| MoCap | – Occlusions |
| | – Constrained place |
| | – Complex calibration and requires expensive equipment |
| RFID | – Objects must be equipped with RFID tags |
| | – The person must wear an RFID |

Motion capture is a sector of research in full evolution. However, the use of such a technology requires a procedure of boring calibration and additional expensive equipment. Furthermore, MoCap has many challenges, for example: occlusion and a constrained space.

In the case of recognizing human actions from the radio-identification RFID, which informs us about the place of the person, RFID labels must equip the objects, which interact with the person, and the port of this sensor by the user is necessary.

A summary of some limitations of different sensors associated with human action recognition is presented in Table 1.

With the progress of microelectronics, human action recognition using wearable inertial sensors, such as the accelerometer or the gyroscope, has been acquiring more and more attentiveness from many researchers. Moreover, the integration of these sensors into different devices, which become a part of people's daily living (such as: smartphones, smart watches, sport medical bracelet, etc.) has opened the way to the advancement of the human action recognition. Among the technologies that recognize human activities, inertial wearable sensors seem to be the most promising. Indeed, their lightweight, small size, and low cost have attracted many researchers (Mimouna et al. 2018). Moreover, the low energy consumption and the reduced computational power provide a long-period recordings and continual interaction compared with based-image processing systems.

Undoubtedly, wearing these sensors is easy and using such a technology can ensure the recognition in darkness. Thanks to all these advantages, the accelerometer, which provides 3-axis accelerations, has been exploited in a diversity of applications in order to detect and analyze human activities.

Furthermore, to enhance the recognition performance, some researchers proposed to combine two different modalities to deal with several realistic events that may appear in the real world, for instance, fusing data from a depth image and data from a wearable inertial sensor as shown in Chen et al. (2015), Malawski and Gałka (2018).

To the best of our knowledge, this is the first research attempt to exploit the potential of the triaxial accelerometer and its employment in various fields especially in HAR. The aim of this chapter is: (i) to present an overview of the state of the art of accelerometers' applications and practicality we focus on the field of HAR exploiting accelerometer data, and (ii) to expose a fusion framework which consists of coupling several information acquired from numerous levels.

As we discussed, several modalities introduced to recognize human activities and as the accelerometer seems to be the most effective; we will present an accelerometer's review and its applications in Sect. 2. The third section is reserved for introducing the field of human action recognition exploiting accelerometers data, in this section we present challenges, various applications related to this field and several approaches employed to guarantee action recognition. Datasets based on inertial sensors are introduced in Sect. 4. We give a detailed description of the fusion framework in Sect. 5. The experimental results are reported in Sect. 6. The seventh section provides the conclusion.

## 2 Accelerometer's Review and Applications

Accelerometers are used to determine the measurement of changes in velocity. There exist two main modes of acceleration measured by this sensor: the first is the linear acceleration, which is the acceleration measured when the change in velocity is in the signal direction, and the second is the centrifugal acceleration, which is the measurement of the displacement of an object in a circle.

The triaxial accelerometer measures the acceleration following three directions $X$, $Y$ and $Z$, as shown in Fig. 1 which represents accelerometer data acquired when moving the phone. It is a kinematic sensor existing in several devices. In addition to game consoles, mobile terminals and automobiles, accelerometers are now present in a large number of connected objects; we mention intelligent textiles, connected watches, cameras, prostheses, shoes, drones, robots, sports and medical bracelets, etc.

Thanks to its many benefits, nowadays, the accelerometer is present in a variety of applications which they will be detailed below.

Recently, monitoring road conditions become necessary to insure safety to vulnerable road users, and also to evaluate the state of the roads. Allouch et al. (2017), developed an android application named RoadSense to predict road conditions using the accelerometer and the gyroscope integrated into the smartphone. According to the results, it guarantees high performance with an accuracy of 98.6%.

In augmented reality, Unuma and Komuro (2015) proposed a natural 3D interaction system, the user can interact with virtual objects superimposed on the real image

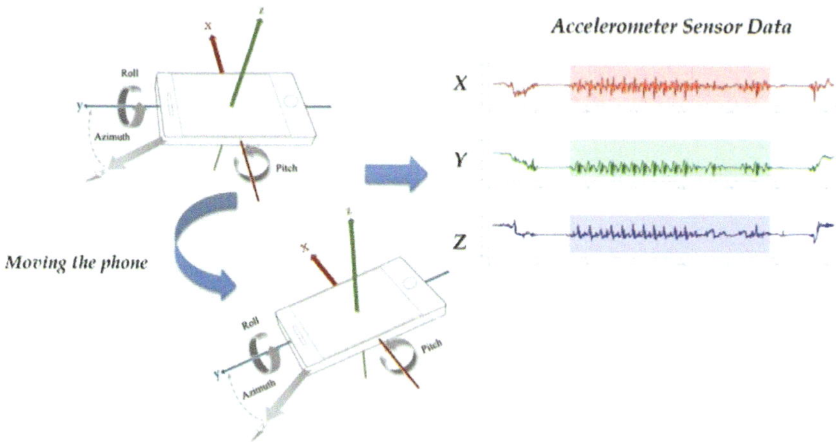


**Fig. 1** Accelerometer sensor data acquired when moving the phone

using his hand. With the aim to insure natural interaction, a triaxial accelerometer is fixed on the depth camera. Thus, when the user pushes a virtual ball, it rolls immediately, and he can just find it when he displaces the mobile display even if the ball quits the screen.

Over the last decade, prosthetics have been evolving owing to the advancement of microelectronics sensors and their facility of incorporation to these prosthetics. In (Beyrouthy et al. 2016), an EEG mind-controlled prosthetic arm is developed. This smart prosthetic arm is controlled through brain commands and it is outfitted with a network of sensors. This smart network provides the prosthetic arm with normal hand movements and intelligent reflexes. Furthermore, the proposed prosthesis has been developed in order to ameliorate the quality of life of patients with a low cost.

In work environments, accelerometers embedded in mobile phones are used for detecting stress levels because it affects the health of workers. Data acquired from the accelerometer was utilised to differentiate humans' behaviours. For 8 weeks, 30 subjects with smartphones from two organizations participated in this study and they noted their stress levels three times while working. Besides, three levels are introduced: low, medium and high stress. An accuracy of 70% for user-specific modal was achieved (Garcia-Ceja et al. 2016).

Also based on a network of sensors embedded in a mobile phone, including the accelerometer and the GPS, Castignani et al. (2015) proposed a new application named SenseFleet, which is capable of detecting risky driving events by identifying several events, such as braking, steering, accelerating and over-speeding. Moreover, the obtained results show that the application is able to precisely identify risky events, it can also differentiate between the drivers' behaviours, for instance calm and aggressive drivers.

Air pollution caused by gaseous emission from vehicles has been increasing with the advancement of economy and vehicles. Traffic conditions are one of the most

**Table 2** A summary of accelerometer's applications

| Refs. | Fields | Applications |
|---|---|---|
| Chen and Shen (2017) | Traffic | – Monitoring roads conditions |
| | | – Detection risky driving events |
| Jain and Kanhangad (2018) | Augmented reality | – The accelerometer provides the tilt of the depth camera to interact with virtual objects |
| Kowalczuk and Merta (2015) | Robotic | – Determination of the position of a mobile robot |
| Kyberd and Poulton (2017) | Medical | – Prosthesis |
| | | – Diagnostic of the heart valve disease |
| Garcia-Ceja et al. (2016) | Healthcare | – Detecting and monitoring stress |
| Zhang et al. (2016) | Pollution | – Detecting driving events to estimate the quality of vehicle emission |
| Rastegari et al. (2017) | Industry | – Machinery maintenance |
| Höflinger et al. (2015) | Biology | – Monitoring movement patterns in animal models of disease |

affecting elements of air pollution, thus, a method based on levels of service is proposed in Zhang et al. (2016), to estimate emissions under various traffic conditions. Accelerometer data was used to describe driving events, which are the characteristics of the vehicle movements that affect the quantity of emission.

In the field of industry, accelerometers are widely used to give an account of the vibration and its changes in the aim of permitting the user to monitor machines, to detect faults and to minimize its suspension. Rastegari et al. (2017) focus on condition based maintenance as regards to machine tools, particularly concentrating on vibration monitoring approaches. Hence, accelerometers are fixed to the spindle units, then, data are transferred to the computer as a dataset in order to be analysed.

A summary of accelerometer's applications is provided in Table 2 in the following.

In conclusion, the accelerometer is exploited in very fields, and is particularly employed to ensure human action recognition, this point will be detailed in the following section.

# 3 HAR Using Accelerometer Data

## *3.1 Challenges*

Although the human action recognition using accelerometers data continues to progress, the recognition accuracy is affected by many challenges in this field. Firstly, people have different motion models: every subject has his unique style of execution as shown in Fig. 2.

Moreover, for the same person, the action may differ from one repetition to another: the action can be shorter or longer as provided in Fig. 3.

Furthermore, the placement of the on-body sensors presents an important defiance, for example: when a person is jogging, the data collected from an accelerometer attached to the wrist is different from data acquired from an accelerometer fixed to the thigh. Figure 4 presents signals recuperate from six different localizations.

In addition, the translation and the rotation of the sensor, when recording the action, may influence the measurement so it may affect the recognition performance. Thus, the number, the position and the type of the accelerometer are principally related to the application. Besides, the complexity of actions and the transition period between two successive actions lead to an additive challenge. Additionally, people performing multiple activities simultaneously might cause confusions.

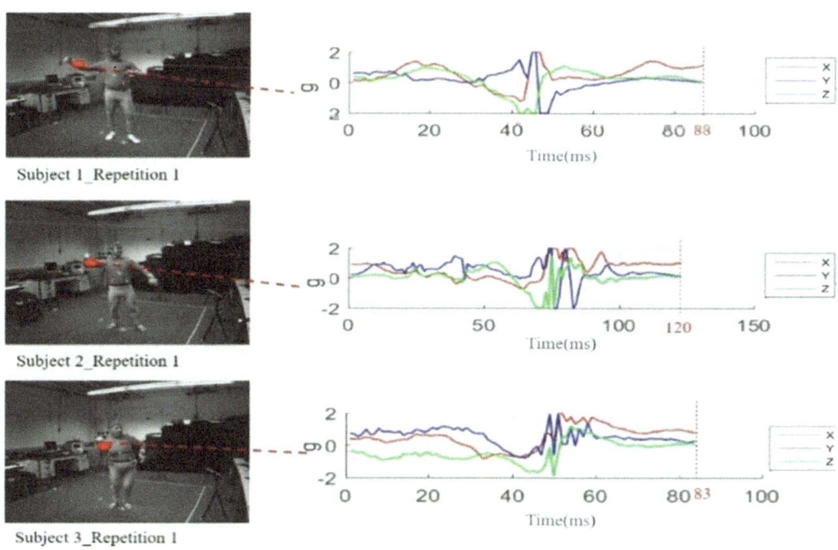


**Fig. 2** Inter-class challenge

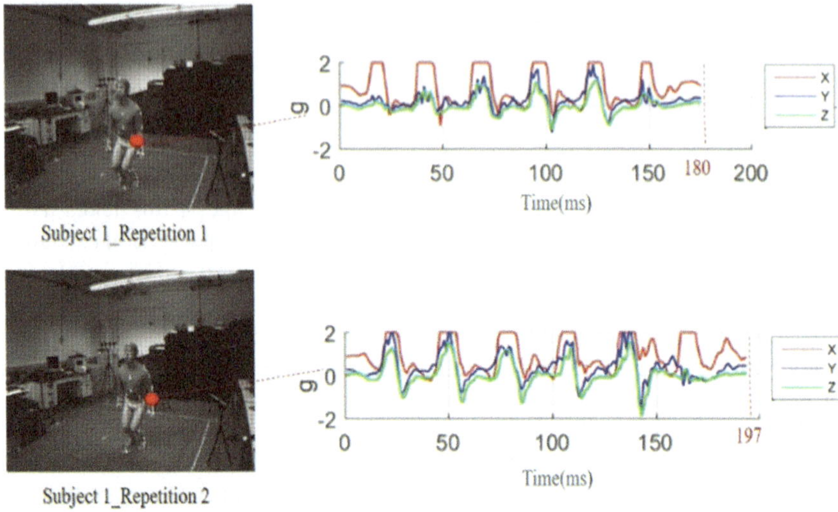


**Fig. 3** Intra-class challenge

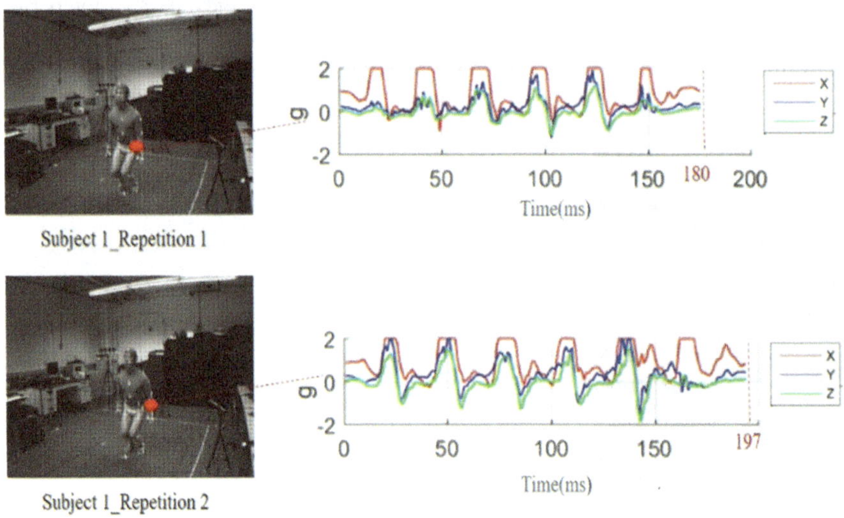


**Fig. 4** Signals acquired from six different positions

## 3.2 HAR Applications

Analysing human actions using wearable sensors, as the accelerometer, has become an increasingly unavoidable area of research in various fields including: medical, virtual reality, sport, security, surveillance, education, etc. In the following, we will expose several applications presented in Table 3 to outline the use of the accelerometer in HAR.

Surgeries are complex tasks accomplished in stressful areas (Zia et al. 2018). Therefore, the immersive virtual reality provides virtual environments to surgeons and trainees to be trained in realistic conditions to ensure the patient's safety and to attenuate errors. Various technologies are used in this field, including wearable sensors, which track the user's motions in order to gain surgical expertise (Dargar et al. 2015).

Laghari et al. (2016) focused on developing a biometric authentication application based on accelerometer data acquired from the smartphone. Indeed, the user performs his signature by handling the phone in his hand and moving it. Ten volunteers participated in this work; each subject had to perform his signature 6 times. The signal

**Table 3** Applications of human action recognition using accelerometer data

| Refs. | Application field | Applications |
|---|---|---|
| Dargar et al. (2015) | Education | – Training in virtual environments to gain expertise |
| Zhang et al. (2015) | Virtual reality | – Training |
| Laghari et al. (2016) | Security | – Biometric authentication application for smartphones |
| | | – Medical adherence |
| Pepa et al. (2015), Kau and Chen (2015), Figueiredo et al. (2016) | Medical | – Diagnostic of various diseases: Parkinson's disease, epilepsy seizure, Alzheimer's disease, etc. |
| | | – Fall accident detection |
| Sen et al. (2015) | Healthcare | – Monitoring sleep |
| | | – Diet monitoring |
| | | – Assessment of elderly people |
| Ferhat et al. | Surveillance | – Assessment of children's behaviours |
| Hung et al. (2015) | Automation | – Controlling the daily environment |
| Hidayat et al. (2016) | Gaming | – Detecting gestures and actions to play |
| Neto et al. (2009) | Industry | – Controlling an industrial robot |
| Koskimäki and Siirtola (2014) | Sport | – Recognizing gym exercises |

matching was used as an identification approach. With regards to the traditional and the graphical techniques, this method is more secure with a false rejection rate of 6.87%.

Kalantarian et al. (2017) proposed an android application implemented on a smartwatch to detect various motions related to medical adherence. Furthermore, the system detects when the bottle is twisted to open it using the accelerometer data and then, the act of revolving the palm to retrieve the pill is identified using gyroscope data. Although the system is sensitive to how to remove the pill, it needed less human involvement for medication adherence with regard to nurses' calls or other forms.

Parkinson's disease is an advancing neurological disorder that affects the basal ganglia. Freezing of Gait (FOG) is one of the most frequent motor disorders for advanced Parkinson disease that can diminish the quality of life and it can be defined as a gait disturbance. Pepa et al. (2015) proposed a smartphone-based application that can detect FOG occurrences and is able to send an acoustic feedback to help patients restore walking. In addition, tested on 18 patients, this method provides an 82.34% of sensitivity.

Kau et al. (2015) used the triaxial accelerometer and the electronic compass integrated in the smartphone, which was located in the pocket of the subject to detect fall accidents. If the system detects a fall event, it will send the user's position identified by the GPS to the rescue center via Wi-Fi or the 3G network. Thus, the user can receive medical help straightaway. An accuracy of 92% is achieved using this algorithm with 450 test actions of 9 types that include a fall event.

Wearable inertial sensors are nowadays used to assist therapeutic movements. In (López et al. 2015), two sensors are worn on the forearm and the upper arm to identify the quality of the patient's movements and observe his/her recovery. Besides, the aim of the study is to define intra and inter-group dissimilarity between a given number of movements accomplished by young people, with regard to motions given by therapists.

Human action recognition is used to analyse children's behaviour and to follow their health and development. Indeed, children's actions can be limited to walking, playing, sitting, sleeping and hand motion. A kindergarten system was developed using acceleration information acquired from the accelerometer fixed on the child's hand, then, these information were anlysed to present a global state of the child's health to parents and child-minders (Kurashima and Suzuki 2015).

The assessment of the elderly people during their daily life became a crucial challenge in order to ensure their safety, autonomy and healthcare. Ferhat et al. focused on recognizing and monitoring elderly people using three inertial units that were mounted on the chest, the right thigh and the left ankle. Additionally, based on real-time technique and data transmission, the subject's motions were continually monitored by healthcare suppliers all along daily activities and abnormal events are detected to intervene.

Over the last decade, home automation has become an important field of research to control the daily environment. In (Hung et al. 2015), a hand gesture recognition belt was developed using an accelerometer and a gyroscope to control a LED array lamp. Indeed, when the user shakes his hand up, the LED turns on and inversely. Consequently, as the user's palm is shaking, the luminosity of the LED can dims.

In the gaming word, the advancement occurs expeditiously. Hidayat et al. (2016) used a Wii remote as a controller of a fighting game. The Wii remote transfers data obtained from the accelerometer that detects gestures or motions of the hand. Then, when the movement is identified, it will be visualized in the built based-Unity 3D game as a player's action.

Neto et al. (2009), developed a system based on two triaxial accelerometers, for the purpose of controlling an industrial robot rather than programing it with typical techniques. Furthermore, the sensors were fixed on the human arms in order to capture its gestures and postures, so the robot can start the movements approximately while the user begins to perform a motion. Besides, a higher performance was achieved using this approach with a recognition rate of 92%.

## 3.3 *Related Work*

Human action recognition using acceleration information has been employed in several application areas mentioned previously; in fact, various approaches described in this section have been proposed to address this challenge.

Pre-processing is considered as a one of the most critical steps that includes replacing missing data or filtering it. Before the feature extraction step, raw data acquired from sensors are generally divided into small segments using windowing technique. In fact, various windowing approaches are used in this level: (i) sliding window that is the most commonly used owing to its facility of implementation and its guaranteed high accuracy, it consists of dividing signals into fixed length windows with or without overlap; (ii) the defined activity windows that resides with the division of the data based on the detection of activity changes; (iii) the defined event windows, where pre-processing is needed to find particular events; (iv) the dynamic sliding window that was developed to overpass the fixed-length of the sliding window technique, the main idea of this novel activity signal segmentation approach is that the window size could be dynamically adapted by using the signal information to determine the most effective segmentation.

Afterwards, feature extraction is considered as a crucial step; which consists of extracting quantities to characterize each performed action. Many researchers tended to extract features commonly from: time domain, frequency domain and time-frequency domain. Time domain characteristics include mean, maximum, median, skewness, variance, etc. Frequency-domain features incorporate peak frequency, signal energy, also the calculation of the power spectral density (PSD) and the utilization of the Fast Fourier Transform (FFT), etc. Furthermore, wavelet transform is the most common technique used to extract features from the time-frequency domain. Adding to this, it exists other techniques employed to extract features from accelerometer signals to differentiate actions such as the Dynamic Time Warping (DTW).

In many works, researchers employed a feature selection process, which consists of selecting a subset of appropriate features from the original features, because the use of inappropriate or redundant characteristics may decrease the performance of the classifier. This process reduces the number of features and the computation time. Generally, it exists three classes in feature selecting: (i) filter methods, (ii) wrapper methods, (iii) hybrid methods. The filter-based method evaluates features without any classifier, so it classes a set of selected features according to the estimated weights of each feature. Different from the filter methods, wrapper, which ensures often the best results, uses classifier accuracies to evaluate the selected subset. Eventually, the hybrid methods consist of selecting the most appropriate features due to some internal parameters of the machine-learning algorithm.

Feature vectors, obtained after extraction/selecting features from raw data, are used in order to train the classification algorithm. Indeed, to ensure this step, many machine learning techniques are employed, which are divided into two principal approaches: supervised and unsupervised methods. In addition, the supervised techniques are based on labeled activity data such as K-nearest neighbours (K-NN), Artificial Neural Networks (ANN-s), Support Vector Machines (SVMs), Decision tree (DT) and Random Forest (RF). Concerning the unsupervised approaches, which are linked with unlabeled data, we can site the Hidden Markov Model (HHM), the K-means, and the Gaussian Mixture Models (GMMs).

Some of the common works introduced to recognize and analyse human actions are presented in the following.

In (López et al. 2015), Lopez et al. proposed a novel method to detect and characterize walking and jogging using a triaxial accelerometer. Actually, the kurtosis of wavelet coefficients or the autocorrelation of the acceleration data was used for the detection. This methodology was tested on three different datasets of walking and jogging.

Lubina et al. (2015) evaluated the application of artificial neural networks (ANNs) to recognize human activities using accelerometer signals. Five accelerometers were fixed on the back, two on the waist laterally and two on the ankles, and 25 subjects were called to perform a set of predefined actions such as sitting down and walking. The obtained signals were firstly filtered using a median filter, then they were partitioned into non-overlapping windows with a length of 0.5s. Afterwards, statistical features were extracted, such as the mean, the sum of squares and the root mean square to train the ANNs. Despite the fact that the implementation of the Fisher Linear Discriminant shows that some features help to discriminate similar actions, none of the axes or the features or the sensors can be neglected.

For monitoring daily life activities, Wang et al. (2016) used a single wearable accelerometer that was attached to the waist and the left ankle respectively with a view to diminish the effect of sensor placement. An ensemble empirical mode decomposition (EEMD), which is a time-analysis technique is introduced in this study. Then, feature selection is insured using a game theory to select relevant features. K-NN and SVM are employed to classify human activities captured from the waist and the ankle. Compared with other works, the results obtained using the proposed method, which selects fewer features, show a better classification.

Monitoring sleep has gained the attention of numerous researchers as it affects our psychological and emotional health. Therefore, Yunyoung et al. (2016) focused on identifying sleep quality based on the triaxial accelerometer and the pressure sensor, and they used various physiological parameters. Additionally, data obtained from the accelerometer determined the sleeping posture and activity. Besides, the proposed algorithm based on a sensor fusion framework effectively detected sleeping and waking situations.

Luštrek et al. (2015) suggested an approach to recognize indispensable lifestyle activities of diabetic patients, using sensors embedded on the smartphone, in order to monitor their lifestyle since it affects the disease. A set of activities was introduced in this study such as eating, sleeping, working, and transport. Five volunteers carried a smartphone and an EEG monitor during two weeks. Furthermore, several features were derived from sensors, such as the user's location, the ambient sound and the acceleration features to train various classifiers to recognize the user's action, such as SVM, RF, and Naïve Bayes. Based on different experiments, the results obtained show that the vote provides a higher accuracy, which combine several machine learning algorithms. To improve the classification rate, they proposed to introduce a final machine learning approach, thus, the accuracy went from 0.77 to 0.88. Nevertheless, it exists some misclassification between the activities such as eating and out.

Noor et al. (2015) proposed a new approach of activity signal segmentation using triaxial accelerometer that consists of a dynamic sliding window. The main aim of this method is to recognize static and dynamic activities as well as transitional activities. Initially, a small window size is adjusted to segment static and dynamic activity signals, then the window length is extended in order to encompass the signal that it can be sometimes longer than the initiated window. Moreover, the dynamic sliding window is used to automatically determinate the optimum window size while the signal is being evaluated. A triaxial accelerometer was fixed on the right waist and three subjects performed several actions such as walking, sitting to lying, standing to sit, etc. and each subject repeated each action five times. For pre-processing, a moving average filter is employed, then a 3s sliding window is used to segment the signal, after that the window length is limited to 1.5s with 50% overlapping rate with the previous window. 117 features are extracted from raw data including standard deviation, spectral entropy, maximum, etc. Afterward, relevant features are selected using Relief-F method. Decision Tree was chosen to classify activities which provided an accuracy of 96% and the transitional activities were effectively recognized.

In (Tran and Phan 2016), sensors integrated on the smartphone were used to develop an android real-time system that is able to recognize human actions. Six actions were introduced such as walking, lying down and sitting. Furthermore, SVM was employed to classify the actions and 248 features were extracted from raw data including mean, minimum, energy, etc. The android system compares the performed activity with its model. Thus an accuracy of 89.59% is achieved using this method. A summary of several approaches introduced for human action recognition using accelerometer data is provided in Table 4.

**Table 4** A summary of various approaches introduced for human action recognition using accelerometer data

| Refs. | Database | Number and placement of the accelerometer | Feature extraction | Feature selection | Classification | Accuracy | |
|---|---|---|---|---|---|---|---|
| Ferhat et al. | 6 subjects<br>12 actions | 3 accelerometers fixed on: the chest, the right thigh and the left ankle | – 11 time domain features: mean, median, peak to peak, variance, etc.<br>– 6 frequency domain features: energy spectrum, energy of the wavelet coefficients,etc.<br>– Fast Fourier Transform<br>– Wavelet analysis | Random Forests<br>– k-NN | k-NN<br>– RF<br>– SVM<br>– HMM<br>– K-means<br>– GMM | : 98%<br>RF<br>SVM<br>HMM<br>K-means<br>GMM | : 98%<br>: 92%<br>: 67%<br>: 51%<br>: 64% |
| Noor et al. (2015) | 3 subjects<br>11 activities | 1 accelerometer mounted on the right waist | – Sliding window technique: initial length is 3s then the size 1.5 s with 50% overlapping<br>– Dynamic sliding window<br>– Time-domain features | Relief-F | – Decision Tree | DT | : 96% |
| Tran and Phan (2016) | 10 subjects<br>6 actions | The phone is placed in the pocket | – Sliding window technique<br>– Features: mean, entropy, etc. | – | – SVM | SVM : | 89.59% |
| Safi et al. (2016) | 6 subjects<br>12 actions | 3 accelerometers placed on: the chest, the right thigh and the left ankle | – The raw data are used as an input for classification algorithm | –<br>–<br>– | – K-means<br>– GMM<br>– HMM<br>– HMMR | K-means:<br>GMM:<br>HMM:<br>HMMR: | 77%<br>70%<br>86%<br>97% |

(continued)

**Table 4** (continued)

| Reference | Database | Number and placement of the accelerometer | Feature extraction | Feature selection | Classification | Accuracy | |
|---|---|---|---|---|---|---|---|
| Chen et al. (2015) | MHAD | 6 accelerometers worn on: the wrists, the ankles and the hips | – Sliding window technique.<br>– Statistical features: rms, variance, mean, std | | $A_1$:86%<br>–SVM<br>$A_5$:56% | $A_2$: 85%<br>$A_3$:71%<br>$A_6$ : 57% | $A_4$ :72% |
| Erdas et al. (2016) | 15 subjects<br>7 activities | The mobile phone is fixed on the chest | – 17 time-domain features: mean. rms, minimum, etc.<br>– 6 frequency-domain features: Fast Fourier Transform<br>– 9 time-frequency domain features: Wavelet Transform | – Chi-square selection<br>– Correlation-based feature selection (CFS)<br>– ReliefF<br>– Gain-ratio-based selection | – RF<br>– k-NN<br>– SVM | RF:<br>K-NN:<br>SVM: | 88%<br>80%<br>67% |
| Zhao et al. (2014) | 23 subjects<br>6 actions | The smart watch is worn on the dominant hand | – The Short-time Fast Fourier Transform using a hamming window is employed | – | – HMM | HMM: | 97% |
| Lee et al. (2017) | 5 subjects<br>3 activities | The smartphone is carried in various locations (held in hand, carried in a bag, placed in a pocket) | – Sliding window technique transformed into a magnitude vector | – Acceleration data<br>– 1D- CNN<br>– RF | 1D-CNN: | 92.71% | |

# 4 Datasets

A large number of public human action recognition datasets have been introduced based on inertial sensors. We distinguish uni-modal and multimodal databases. This section consists of a review of various databases that have been included to recognize human actions captured from accelerometer data.

## *4.1 Uni-Modal Databases*

### 4.1.1 MIT PlaceLab Dataset

This Dataset is one of the first public databases in this field of research. To record this dataset, five accelerometers and a wireless heart rate monitor were utilised, each accelerometer is mounted on the left and right arm, the left and right leg and one on the hip. During a four-hour period, one person is asked to perform a set of activities wearing these sensors, including house-holding activities, such as preparing a recipe, cleaning the kitchen, doing the laundry and other types of everyday tasks, for instance talking to the phone or answering emails. However, data existing in this database are collected from one person, which could present a real problem because each person has its own way to perform activities, so the characteristics of the action are poorly represented.

### 4.1.2 UC Berkeley WARD Dataset

WARD (Wearable action recognition database) is a public human action recognition dataset developed by the University of California. It consists of continuous sequences of human actions measured by a network of wearable motion sensors. The sensors are attached at five body locations: the two wrists, the waist, and the two ankles. Each wireless sensor includes a triaxial accelerometer and a biaxial gyroscope. The database contains 20 subjects: 13 male and 7 female and includes a rich set of activities that involve some of the most frequent actions in the daily life, such as standing, sitting, walking and jumping. It is true that WARD covers the most typical human actions and includes a sufficient number of persons, but some of the data is missed due to battery failure.

### 4.1.3 USC-HAD

A single inertial sensor was used to evaluate 12 different actions performed by 14 subjects (7 males and 7 females): each action is repeated four times. This database includes a considerable number of subjects of different sexes and the activities considered are among the most basic and common human activities in people's daily lives. However, data is acquired form a single accelerometer

#### 4.1.4 REALDISP (REAListic Sensor DISPlacement)

Realistic sensor displacement is a benchmark dataset dedicated for human action recognition. This set was collected to evaluate the effects of sensor displacement in activities recognition, "which can be caused by a loose fitting of sensors, or a displacement by the users themselves". Indeed, three scenarios were introduced: ideal-placement, self-placement, and induced-displacement. The first scenario is "Ideal placement" or default scenario, where sensors are arranged by the instructor on predefined locations of the body. The second scenario is the "Self-placement", where the user is asked to position 3 sensors himself on the body part specified by the instructor. This scenario tries to simulate some of the variability that may occur in the day-to-day usage of an activity recognition system, involving wearable or self-attached sensors. And for the last scenario, the instructor introduces a de-positioning of the sensors using rotations and translations with respect to the ideal placement. This database consists of 33 different physical activities that can be classified as warming up, cooling down and fitness activities and it includes 17 subjects. Data was measured from nine different sensors that contain a 3D accelerometer; a 3D gyroscope, a 3D magnetic field orientation and a 4D quaternion that are attached overall body parts.

Table 5 lists a summary of some uni modal publicly available databases using accelerometers for human action recognition.

### *4.2 Multimodal Databases*

#### 4.2.1 CMU Multimodal Activity Database

This Database was developed in the Carnegie Mellon University that contains different multimodal measures of the human activity of subjects performing the tasks involved in cooking and food preparation. It contains video, audio, RFID tags and motion capture system based on-body markers and physiological sensors such as galvanic skin response (GSR) and skin temperature. In addition, 43 subjects were asked to perform food preparation and cook five recipes while the sensors were placed all over the body: both forearms and upper arms, left and right calves and thighs, abdomen, and both wrists. This set involves a very large population but it is specific to just cooking activities.

**Table 5** Summary of uni-modal publicly available databases using accelerometer data for human action recognition. $N_s$: Number of Subjects. $N_A$: Number of Accelerometers

| Database | $N_s$ | Actions | $N_A$ | Sensor locations | Comments |
|---|---|---|---|---|---|
| MIT Place Lab | 1 | 10 activities: Prepare a recipe Do a load of dishes Search for items, etc. | 5 | Right arm, Left arm Right leg, Left leg Hip | – Just one person performing the activities – Activities considered are imprecise |
| USC Berkeley WARD | 20 | 13 activities: Stand Sit Walk Forward Walk left-cercle, etc. | 5 | Left wrist, Right wrist, Front centre of the waist Left ankle, Right ankle | – Part of the data is missing |
| USC-HAD | 14 | 12 activities: Walk forward Walk left Walk right Walk up stairs, etc. | 1 | Front right hip | – Data collected form a single accelerometer |
| Realistic sensor displacement benchmark dataset (Realdisp) | 17 | 33 activities: Walking Jogging Running Jump up Trunk twist, etc. | 9 | Left calf, Left thigh Right calf, Right thigh Back, Left lower arm Left upper arm, Right lower arm, Right upper arm | – The dataset covers a wide range of physical activities and a number of wearable sensors |

### 4.2.2 OPPORTUNITY Dataset

The opportunity dataset is collected from a European research project called OPPORTUNITY, which concentrated on daily home activities especially on preparing breakfast. This dataset includes different modalities such as accelerometers, gyroscopes, magnetometers, microphones, and cameras. 12 subjects were asked to perform a sequence of daily morning activities including grooming a room, preparing and drinking coffee. Different modalities were used to collect data, such as a camera, a microphone, an accelerometer, and a gyroscope.

### 4.2.3 Berkeley MHAD: Multimodal Human Action Database

MHAD contains temporally synchronized and geometrically calibrated data acquired from an optical motion capture system, multi stereo cameras from multiple views, depth sensors, accelerometers and microphones. 11 subjects (7male and 7 female) participated in the data collection and were asked to perform 11 actions with five repetitions for each action, including jumping in place, jumping jacks, bending, waving two hands. Prior to each recording, the subjects were given instructions on what action to perform; however, no specific details were given on how the action should be executed (i.e., performance style or speed). In addition, six accelerometers were fixed on the wrists, ankles and hips, and the two Kinect were placed in opposite directions. This database contains 660 action sequences.

### 4.2.4 UTD-MHAD: University of Texas at Dallas Multimodal Human Action Dataset

UTD-MHAD is a publicly available multimodal human action recognition data set collected from a Kinect and a wearable inertial sensor measuring a 3 axis : accelerometer, velocity signals and magnetic strength. The dataset contains 8 subjects (4 female and 4 male) and 27 different actions: right arm swiping to the left, right arm swiping to the right, right hand waving, two-hand clapping, right arm throwing, crossing the arms, etc. Each person repeats each action four times with the wearable inertial sensor fixed on the subject's right wrist or right thigh depending on whether the action was mostly an arm or a leg type of action.

### 4.2.5 Huawei/3DLife Dataset

The Huawei/3DLife is a multimodal dataset developed for a 3D human reconstruction and action recognition Grand Challenge in 2013. For this challenge, two datasets were provided: Dataset 1 contains a synchronized RGB-plus-Depth video captured by five Kinects, as well as multiple-Kinects audio and eight inertial sensors covering the whole body. The inertial sensors were placed on: the left wrist, the right wrist, the chest, the hips, the right ankle, the left ankle, the right foot and the left foot. This dataset includes two sessions with different spatial arrangements of the sensors. 17 subjects performed a set of 22 repetitive actions, and each action was performed 5 times. It consists approximately 3740 captured gestures. The performed actions can be classified into i) Simple actions that involve mainly the upper human body, ii) Training exercises, iii) Sports related activities and iv) Static gestures.

With regard to Dataset 2, it was captured in Berlin and includes synchronized multi-view HD video streams of multiple humans doing multiple actions. It consists of 7 individuals performing a set of 26 different body movements.

#### 4.2.6 Multimodal Kinect-IMU dataset

This dataset has been originally collected to investigate transfer learning among ambient sensing and wearable sensing systems. Nevertheless, the dataset may be also used for gesture spotting and continuous activity recognition. It includes data for three activity recognition scenarios, namely HCI (gesture recognition), fitness (continuous recognition) and background (unrelated events). It comprised synchronized 3D coordinates of 15 body joints, measured by a vision-based skeleton tracking system (Microsoft Kinect), and the readings of 5 body-worn inertial measurement units (IMUs. A single subject performs five kinds of geometric gestures with the right hand in alternation 48 times. The locations of the IMUs devices are: the left lower arm, the right lower arm, the back, the left upper arm and the right upper arm.

Table 6 lists a summary of some multi-modal publicly available databases involving accelerometer sensor for human action recognition..

## 5 Fusion Framework

Although human action recognition promises to be highly effective, the exploitation of multi-level fusion approaches can guarantee an excellent rate thanks to the wealth of the information available in all stages of the human action recognition process: acquisition, feature extraction, classification and decision. Thus, we introduce a fusion framework that utilises accelerometers data.

Fusing data is the process of coupling data acquired from numerous sources (in our case several accelerometers) allowing to assess the accuracy of the system. Indeed, we distinguish two categories of merging: before correspondence and after correspondence. The first category concerns the signal-level fusion and the feature level fusion, and the second category involves fusion at the score level and fusion at the decision level. The four levels of fusion shown in Fig. 5, are presented in the following.

### *5.1 Signal-Level Fusion*

The signal presents the modality acquired on-line or off-line (ex. Speech, Accelerometer signal, Image, Video, etc.). At this level, the fusion is only possible when the data are compatible: the sources produce signals of the same type. In our study, the signal fusion technique includes the combination of 3-axes signals from the accelerometer ($X$-axis, $Y$-axis and $Z$-axis).

**Table 6** Summary of multimodal publicly available databases using accelerometer data for human action recognition. $N_s$: Number of Subjects

| Database | $N_S$ | Actions | Sensors | Accelerometers locations | Comments |
|---|---|---|---|---|---|
| CMU Database | 43 | Food preparation Cooking five recipes | Camera/ Microphone Motion capture Accelerometer, etc. | Both forearms, Upper arms, Left/ Right calves, Thighs, Abdomen, Both wrists | – This set is concentrated on a very specific category of actions (cooking) |
| Opportunity | 12 | Groom room, Prepare coffee, Clean up, etc. | Accelerometer/ Gyroscope Microphone/ Camera, etc. | Wrist, Chest Limb, Shoulder Foot | – The database covers only morning activities |
| Berkeley MHAD | 11 | 11 actions : Jumping in place, Jumping jacks, Bending, Punching, etc. | 12 cameras/ 2 Kinect 6 accelerome-ters/ motion capture/4 microphones | Wrists Ankles Hips | – Data are temporally synchronized and controlled |
| UTD-MHAD | 8 | 27 actions : jogging in place, walking in place, sit to stand, etc. | – Kinect Wearable inertial sensor | The right wrist or the right thigh | – The dataset covers a wide range of actions |
| Huawei/3DLife: Dataset 1 | 17 | 22 actions : Throwing, Jumping Jacks Squats, etc. | 5 Kinect -8 inertial sensors | Right/Left wrist Chest, hips, Right/Left ankle, Right/Left foot | – This set focuses on various types of actions |
| Multimodal Kinect-IMU dataset | 1 | Five kinds of geometric gestures :Drawing a triangle, etc. | – Kinect -5 inertial sensors | Left /Right lower arm Back Left/ Right upper arm | – Just one person performed the gestures |

## 5.2 *Feature Level Fusion*

Features or attributes are characteristics extracted from the raw data. The feature fusion level is the combination of the different feature vectors, obtained either from the same modality or from different modalities. Therefore, the merging at this level can consider homogeneous feature vectors and heterogeneous feature vectors.

**Table 7** Score level fusion rules. $T$ is the number of matchers and $s_j$ presents the normalized scores of the $j^{th}$ matcher. $w_j$ corresponds to the Equal error rate (EER) of the $j^{th}$ and $F$ represents the fusion score

| Rule | Equations |
|---|---|
| Sum | $F = \frac{1}{T}\sum_{j=1}^{T} s_j$ |
| Max | $F = \max(s_1, s_2 ..., s_T)$ |
| Min | $F = \min(s_1, s_2 ..., s_T)$ |
| Product | $F = \prod_{j=1}^{T} s_j$ |
| Weighting | $F = \sum_{j=1}^{T} w_j s_j$ |

## 5.3 Score Level Fusion

A score is a measure of similarity that corresponds to the distance between the test sample and the reference sample. In fact, the fusion at this level presents a compromise between the richness of the information and the facility of the implementation. Actually, each classifier produces a matching score or several scores and the merging process combines these measures to obtain the final score which will be then used to produce the final decision. There exist two main approaches to combine scores: the classification of scores and the combination of scores. Several rules exploited to ensure the fusion of scores are presented in Table 7.

## 5.4 Decision Level Fusion

It processes the outputs of the different classifiers. The decision level fusion consists in assembling the decisions obtained from each classifier in order to obtain the final decision. There are several methods for merging decisions, such as the AND and the OR logic operator type rules, the majority vote, and the Dempster–Shafer theory.

Thus, we focused our attentiveness on characterizing the human actions in order to gain a better classification accuracy by employing a fusion framework exploiting every information obtained along the human action recognition process.

For the feature extraction, we opted for the discrete wavelet transform. The Wavelet transform of a function $f(x)$ is calculated using (1) as follows:

$$W_f(i, \tau) = \int_{-\infty}^{+\infty} f(x)\psi_{i,\tau}^{\star}(x)dx \tag{1}$$

$$\psi_{i,\tau}(x) = \frac{1}{2^i}\psi\left(\frac{x-\tau}{i}\right) \tag{2}$$

$\psi$ is the wavelet mother.

The wavelet transform decomposes initially raw data into approximation coefficients by employing low pass filter and detail coefficients by a high-pass filter. Various levels are constructed as follows: the approximation signal required from the previous level is decomposed into approximation and detail coefficients. The desired decomposition level is determined after several repetitions of this process.

With regard to the classification, we exploit support vector machine SVM.

The experimentations associated to the fusion framework are presented in the following section.

# 6 Experimental Results and Analysis

## 6.1 Results

To evaluate the effectiveness of our methodology, we chose three databases: WARD, MHAD and Realdisp which were detailed in Sect. 5. The position of each accelerometer related to each dataset is presented in Fig. 6.

In the interest of the fusion framework, we aimed firstly to select the sensors that guarantee better classification rates for each dataset. In fact, these sensors will be afterwards exploited in the fusion approach in order to provide a higher performance.

Therefore, we proposed to evaluate each accelerometer individually. The collected data from the accelerometer sensor were firstly divided into N temporal windows using the sliding window technique. The window length related to each database is 6 for WARD, 15 for MHAD and 9 for Realdisp. The number of segments N was

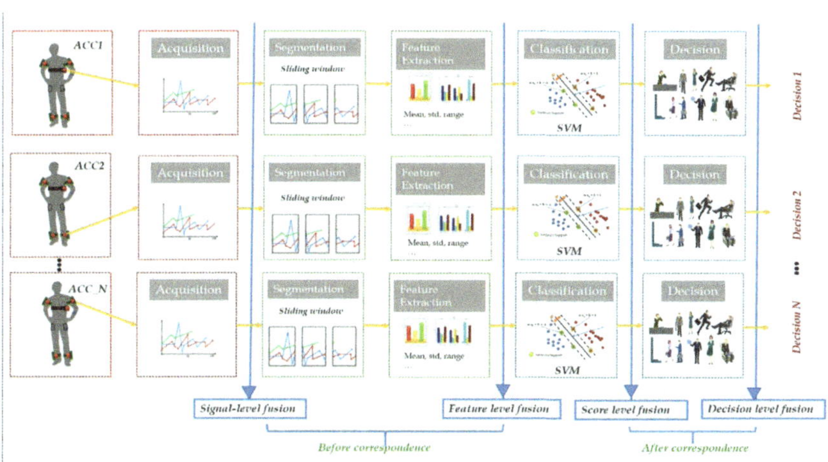


**Fig. 5** An example of multi-level fusion using accelerometer data

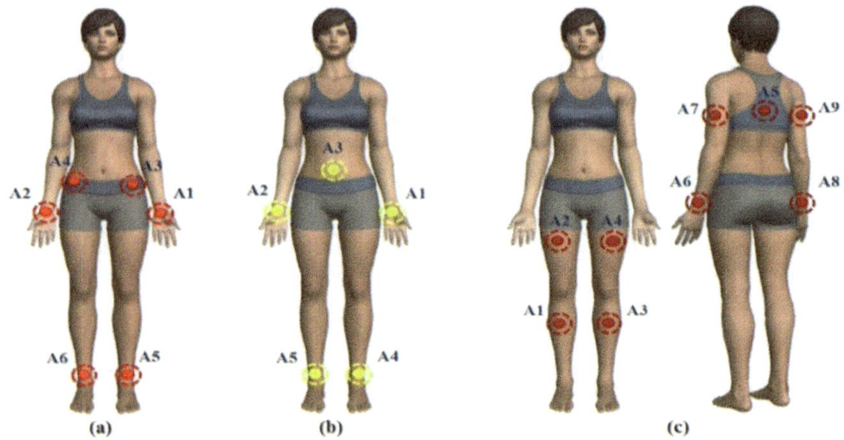


**Fig. 6** Positions of the accelerometers for the three databases. **a** MHAD. **b** WARD. **c** Realdisp

**Table 8** Recognition rates (%) for all accelerometers for the 3 databases

| Accelerometer | Recognition rate (%) | | |
|---|---|---|---|
| | WARD | MHAD | Realdisp |
| $A_1$ | 64 | 88 | 41 |
| $A_2$ | 64 | 86 | 49 |
| $A_3$ | 54 | 36 | 30 |
| $A_4$ | 72 | 68 | 50 |
| $A_5$ | 71 | 45 | 41 |
| $A_6$ | – | 50 | 51 |
| $A_7$ | – | – | 51 |
| $A_8$ | – | – | 43 |
| $A_9$ | – | – | 48 |

determined based on several experimentations. Then, the features were extracted from each window using the discrete wavelet transform with a Daubechies2 as a wavelet mother. From the approximation coefficients, we extracted the mean and the standard deviation and concerning the detail coefficients, we extracted the minimum and the root mean square. These measures are computed over the three directions ($X$, $Y$ and $Z$) within each temporal window. Furthermore, we took advantage of the SVM with RBF kernel to classify the actions. We considered 12 subjects for training and 8 subjects for the test. Concerning MHAD, 7 subjects were reserved for training and 5 persons for testing. Finally, for Realdisp 10 subjects were provided for the learning base and 7 persons were preserved for testing.

Thus, the results relative to each accelerometer of the three datasets are presented in Table 8.

**Table 9** Recognition rates (%) for all levels of fusion

| Database | Fusion level | | | | | | |
|---|---|---|---|---|---|---|---|
| | Signal | Feature | Score | | | Decision | |
| | | | Sum | Max | Product | AND | Vote |
| WARD | 75 | 88 | 88 | 82 | 88 | 72 | 78 |
| MHAD | 89 | 94 | 94 | 91 | 97 | 86 | 89 |
| Realdisp | 53 | 76 | 77 | 67 | 78 | 51 | 63 |

According to the results shown in Table 8, we can observe that for the WARD database the accelerometers attached to the ankles ($A_4$ & $A_5$) provide a better performance in terms of accuracy. Indeed, the actions introduced in this dataset are essentially linked with the motions of the feet such as "walking", "going up and down the stairs", "Jumping", etc. Accordingly, the sensors fixed on the ankles ensure better recognition rates compared to the other sensors.

Form the results obtained from the six accelerometers considered on the MHAD dataset, we notice that the accelerometers mounted on the left and the right wrist ensure a better classification rates ($A_1$ & $A_2$). Based on the type of the actions considered in this dataset, which are related to the hand motions (e.g. "clapping", "waving", "punching"), the accelerometers worn on the hands can classify correctly the classes. With regard to the accelerometers attached to the ankles, they are not able to generate useful information because the actions are relatively static.

Regarding the Realdisp dataset, the accelerometers $A_4$, $A_6$ and $A_7$ attached respectively to the right thigh, the left lower arm and the left upper arm, seem to be the most effective to distinguish the human actions introduced in this dataset. In fact, the actions employed focus on the trunk, upper and lower extremities including actions of translation, jumping and physical activities. Therefore, a part or all of the body is moving during the performance of the actions; otherwise, the recognition rates relative to the 9 accelerometers distributed at different positions are convergent.

After evaluating each sensor separately and with a view to obtain a higher recognition rate and improve the classification, we proposed to employ the multi-level fusion techniques. We fused the signals acquired from the 3 axis acceleration data and we combined the features from the time-frequency domain from each chosen sensor. For the score level, the Sum, the Max and the Product rules were exploited. Finally, for the decision level, the rules AND and OR were employed.

In this step, we suggest involving the 3 accelerometers that guarantee the best performance for each dataset based on the positions of the sensors and the results obtained. For WARD database accelerometer number 1, accelerometer number 4 and accelerometer number 5 are chosen. In addition, $A_1$, $A_2$ and $A_4$ are part of the stage of fusion for MHAD. And for Realdisp, only $A_4$, $A_6$ and $A_7$ are involved. Thus, the results are presented in Table 9.

## 6.2 Discussion

We compared the recognition rates of the multi-level fusion framework with the accuracies obtained from each accelerometer individually, we noticed that combining signals, features, scores or decisions guarantees a higher performance. In fact, our approach exploits every information available in the recognition process from acquisition to decision and leads to good results for the employed datasets as we listed in Table 9.

From Table 9, we outline that the matching score level fusion outperformed the other levels of fusion and achieved favorable performance compared with the utilization of each sensor individually. Actually, compared with the other levels of coupling, this level provides richer information as it fuses the distances between the test samples and the reference samples.

Moreover, the classification accuracy using fusion scores is higher than the performance found in the literature for MHAD database which is 97% against 94% in Chen et al. (2015).

This improvement leads to a discrimination between most of the actions as we can see from Figs. 8 and 9 which represent the confusion matrices when fusing scores related respectively to MHAD database and WARD database

In the intention to evaluate the effectiveness of our method, we compare the confusion matrix related to the use of each accelerometer individually with the matrix obtained from the coupling of scores. Thus, we consider MHAD database as an example, Fig. 7a, b and c correspond respectively to the confusion matrix of $A_1$, $A_2$ and $A_4$.

As seen in Fig. 7a, the accelerometer $A_1$ worn on the left wrist provides a good discrimination between the actions as the accomplishment of most of the actions requires the contribution of the left hand differently (waving, punching, throwing, etc.). However, it can't differentiate action 4 "Boxing" from action 7 "Clapping" owing to the similarity of the behavior of arms. In addition, the misclassification that occurs between action 8 "Throwing" and action 11 "Standing" can be explained by the fact that the posture of the left hand is the same in these actions so the accelerometer generates similar raw data.

From Fig. 7b, we notice that the distinction between action 5 and action 6 is difficult using the accelerometer $A_2$ fixed on the right, besides, the action 6 "waving using the right hand" can be considered as a subset of the action 5 "waving using both hands".

Figure 7c shows the confusion matrix when using the accelerometer $A_4$ mounted on the right hip, we notice that the recognition of classes: 9, 10 and 11 is improved by this sensor because of its contribution to the accomplishment of these tasks: To stand up then sit, Sit and Stand up. However, the system thus is unable to distinguish between the other classes.

As seen in Fig. 8, combing the scores acquired from these sensors leads to a discrimination between most of the actions, nonetheless, there remain some slight

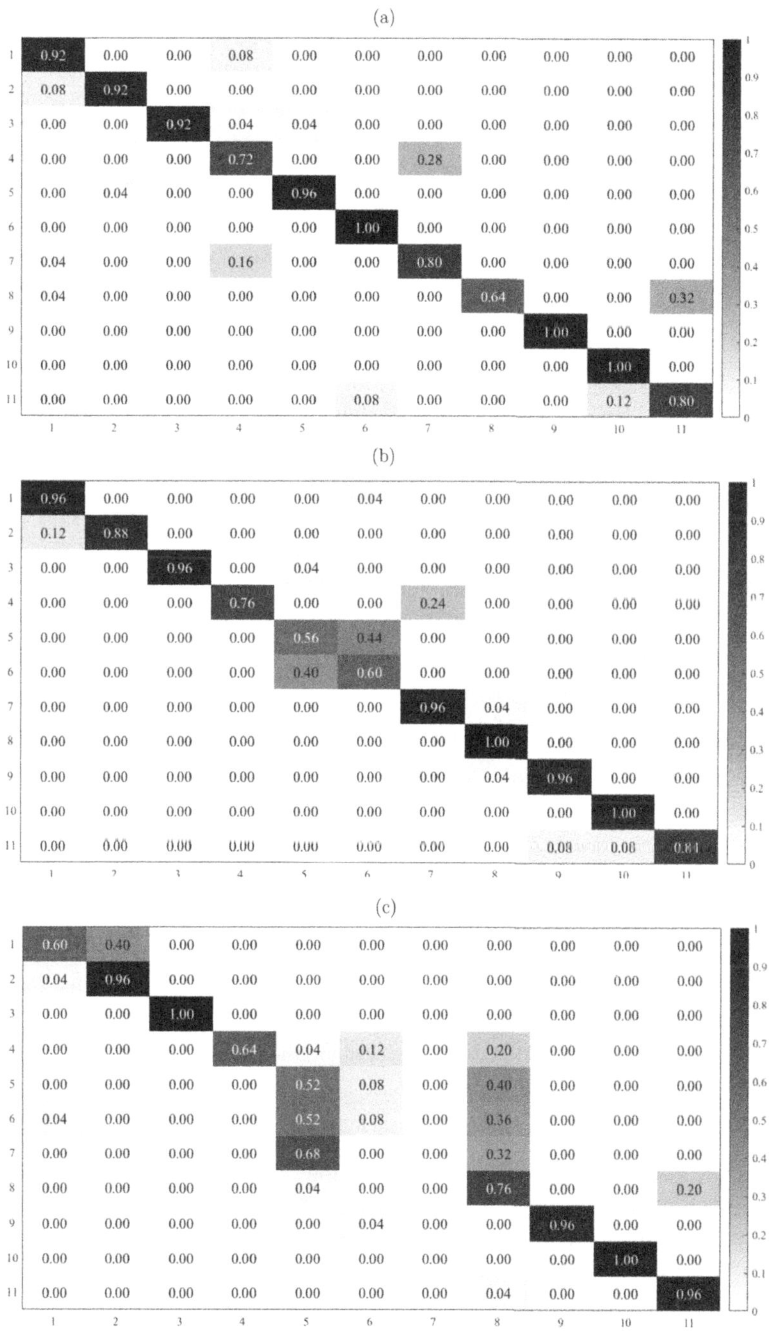


**Fig. 7** Confusion Matrix related to MHAD **a** when using $A_1$ **b** when using $A_2$ **c** when using $A_4$ (Actions: 1. Jump, 2. Jack, 3. Bend, 4. Punch, 5. Wave 2 hands, 6. Wave using the right hand, 7. Clap 8. Throw 9. Sit+Stand, 10. Sit, 11. Stand)

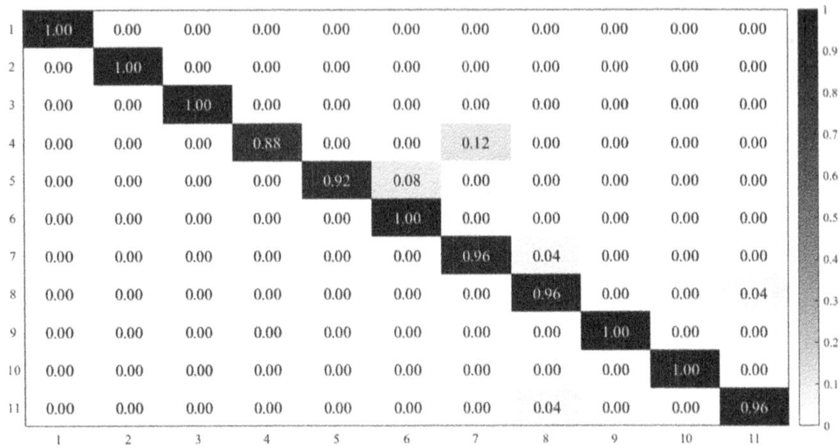


**Fig. 8** Confusion matrix of MHAD database related to fusion scores for $A_1$ & $A_2$ & $A_4$ (rule: Product)

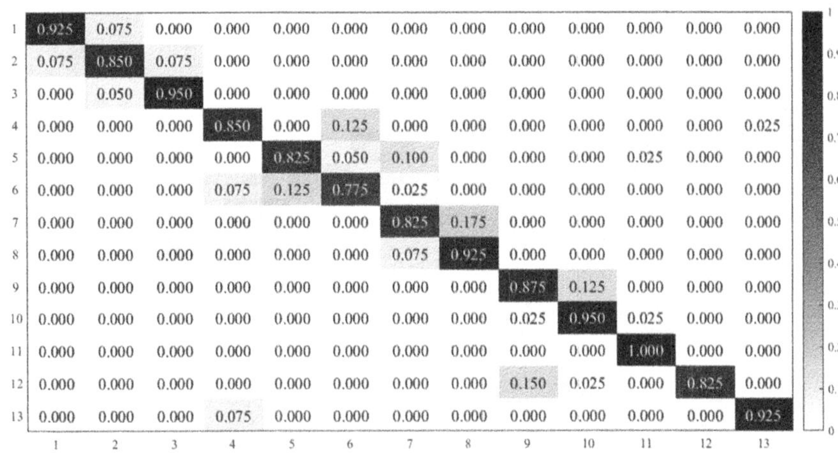


**Fig. 9** Confusion matrix of WARD database related to fusion scores for $A_1$ & $A_4$ & $A_5$ (rule: Product)

misclassifications between the action "punching" and "clapping" because of the similarity of hand movements.

As regards to WARD database, the misclassification occurs between the most similar actions as "walk forward", "walk right" or "walk left" as shown in Fig. 9. Indeed, the walking speed differs from one person to another so the differentiation of these actions is a challenging task.

Finally, the classification accuracies of our method are encouraging as it decreases the number of misclassifications and provides important recognition rates.

# 7 Conclusions

In this chapter, an overview of different methodologies for human action recognition using accelerometer data have been introduced. After we covered diverse sensors used to recognize human actions, we proved that the accelerometer seems to be the most efficient thanks to its benefits in this area of research. Furthermore, various applications related to human action recognition in many areas were outlined, and different approaches existing in the literature were reviewed. Moreover, we reported some publicly available datasets from human action recognition where the accelerometer data was provided. Afterward, a multi-level fusion framework was introduced using acceleration data from the most efficient accelerometers for each dataset used to evaluate this work. The multi-level fusion framework included a signal level, a feature level, a score level and a decision level. According to the results, the recognition rates were improved however; there remains some slight misclassification between the most similar classes.

# References

Aggarwal, J., & Ryoo, M. (2011). Human activity analysis: A review. *ACM Computing Surveys, 43*(3), 1–47.

Allouch, A., Koubaa, A., Abbes, T., & Ammar, A. (2017). RoadSense: Smartphone application to estimate road conditions using accelerometer and gyroscope. *IEEE Sensors Journal, 17*(13), 4231–4238.

Ameur, S., Ben Khalifa, A., & Bouhlel, M.S. (2016). A comprehensive leap motion database for hand gesture recognition. In *7th International Conference on Sciences of Electronics, Technologies of Information and Telecommunications* (pp. 514–519).

Beyrouthy, T., Kork, S. K. A., Korbane, J. A., & Abdulmonem, A. (2016). EEG mind controlled smart prosthetic arm. In *IEEE International Conference on Emerging Technologies and Innovative Business Practices for the Transformation of Societies, EmergiTech* (pp. 404–409).

Castignani, G., Derrmann, T., Frank, R., & Engel, T. (2015). Driver behavior profiling using smartphones: A low-cost platform for driver monitoring. *IEEE Intelligent Transportation Systems Magazine, 7*(1), 91–102.

Chebli, K., & Ben Khalifa, A. (2018). Pedestrian detection based on background compensation with block-matching algorithm. In *15th International Multi-Conference on Systems, Signals & Devices (SSD'18)* (pp. 473–477).

Chen, C., Jafari, R., & Kehtarnavaz, N. (2015). Improving human action recognition using fusion of depth camera and inertial sensors. *IEEE Transactions on Human-Machine Systems, 45*(1), 51–61.

Chen, Y., & Shen, C. (2017). Performance analysis of smartphone-sensor behavior for human activity recognition. *IEEE Access, 5*, 3095–3110.

Cornacchia, M., Ozcan, K., Zheng, Y., & Velipasalar, S. (2017). A survey on activity detection and classification using wearable sensors. *IEEE Sensors Journal, 17*(2), 386–403.

Dargar, S. Kennedy, R., Lai, W., Arikatla, V., & De, S. (2015). Towards immersive virtual reality, (iVR): A route to surgical expertise. *Journal of Computational Surgery, 2*(2), 1–26.

Erdaş, Ç., Atasoy, I., Açıçı, K., & Oğu, H. (2016). Integrating features for accelerometer-based activity recognition. *Procedia Computer Science, 98*, 522–527.

Figueiredo, I. N., Leal, C., Bolito, L. P., & Lemos, A. (2016). Exploring smartphone sensors for fall detection. *MUX: The Journal of Mobile User Experience*, *5*(2), 1–17.
Garcia-Ceja, E., Osmani, V., & Mayora, O. (2016). Automatic stress detection in working environments from smartphones' accelerometer data: A first step. *IEEE Journal of Biomedical and Health Informatics*, *20*(4), 1053–1060.
Hidayat, A. A., Wasista, S., & Pratiwi, Y. P. (2016). Development of fighting genre game(boxing) using an accelerometer sensor. In*International Conference on Knowledge Creation and Intelligent Computing* (pp. 201–206).
Höflinger, F. Rui, Z., Tobias, V., Enrique, G., Adnan, Y., Christina, S., Kerstin, K., & Leonhard, M. (2015). Motion capture sensor to monitor movement patterns in animal models of disease. In *IEEE 6th Latin American Symposium on Circuits & Systems* (pp. 1–4).
Hung, C.-H., Bai, Y.-W., & Wu, H.-Y. (2015). Home appliance control by a hand gesture recognition belt in LED array lamp case. In *IEEE 4th Global Conference on Consumer Electronics, GCCE* (pp. 599–600).
Jain, A., & Kanhangad, V. (2018). Human activity classification in smartphones using accelerometer and gyroscope sensors. *IEEE Sensors Journal*, *18*(3), 1169–1177.
Jegham, I., & Ben Khalifa, A. (2017). Pedestrian detection in poor weather conditions using moving camera. In *IEEE/ACS 14th International Conference on Computer Systems and Applications, AICCSA* (pp. 358–362).
Jegham, I., Ben Khalifa, A., Alouani, I., & Mahjoub, M. A. (2019). MDAD: A multimodal and multiview in-vehicle driver action dataset. In M. Vento & G. Percannella (Eds.), *Computer analysis of images and patterns, CAIP, 2019* (Vol. 11679, pp. 518–529). Lecture notes in computer science. Cham: Springer.
Kalantarian, H., Alshurafa, N., & Sarrafzadeh, M. (2016). Detection of gestures associated with medication adherence using smartwatch-based inertial sensors. *IEEE Sensors Journal*, *16*(4), 1054–1061.
Kau, L.-J., & Chen, C.-S. (2015). A smart phone-based pocket fall accident detection, positioning, and rescue system. *IEEE Journal of Biomedical and Health Informatics*, *19*(1), 44–56.
Koskimäki, H., & Siirtola, P. (2014). Recognizing gym exercises using acceleration data from wearable sensors. In *IEEE Symposium on Computational Intelligence and Data Mining* (pp. 321–328).
Kowalczuk, Z., & Merta, T. (2015). Evaluating the position of a mobile robot using accelerometer data. In *Advanced and intelligent computations in diagnosis and control* (pp. 131–143).
Kurashima, S., & Suzuki, S. (2015). Improvement of activity recognition for child growth monitoring system at kindergarten. In *41st Annual Conference of the IEEE Industrial Electronics Society* (pp. 2596–2601).
Kwon, B., Junghwan, K., Kyoungoh, L., Yang, K., Sangjoon, P., & Sanghoon, L. (2017). Implementation of a virtual training simulator based on 360° multi-view human action recognition. *IEEE Access*, *5*, 12496–12511.
Kyberd, P., & Poulton, A. (2017). Use of accelerometers in the control of practical prosthetic arms. *IEEE Transactions on Neural Systems and Rehabilitation Engineering*, *25*(10), 1884-1891.
Laghari, A., Waheed-ur-Rehman, & Memon, Z. A. (2016). Biometric authentication technique using smartphone sensor. In *13th International Bhurban Conference on Applied Sciences and Technology* (pp. 381–384).
Lee, S.-M., Yoon, S. M., & Cho, H. (2017). Human activity recognition from accelerometer data using convolutional neural network. In *IEEE International Conference on Big Data and Smart Computing* (pp. 131–134).
Lejmi, W., Ben Khalifa, A., & Mahjoub, M. A. (2017). Fusion strategies for recognition of violence actions. In *IEEE/ACS 14th International Conference on Computer Systems and Applications, Hammamet* (pp. 178–183).
Lejmi, W., Ben Khalifa, A., & Mahjoub, M. A. (2019). Challenges and methods of violence detection in surveillance video: A survey. In M. Vento, G. Percannella (Eds.), *Computer analysis of images*

*and patterns. CAIP, 2019* (Vol. 11679, pp. 62–73). Lecture notes in computer science. Cham: Springer.

López, J. D., Sucerquia, A., Duque-Muñoz, L., & Vargas-Bonilla, F. (2015). Walk and jog characterization using a triaxial accelerometer. In *IEEE International Conference on Computer and Information Technology; Ubiquitous Computing and Communications; Dependable, Autonomic and Secure Computing; Pervasive Intelligence and Computing* (pp. 1406–1410).

Lubina, P., & Rudzki, M. (2015). Artificial neural networks in accelerometer-based human activity recognition. In *22nd International Conference Mixed Design of Integrated Circuits & Systems* (pp. 63–68).

Luštrek, M., Mitja, L., Cvetkovi'c, B., Mirchevska, V., Stefan, J., Kafali, Ö., Romero, A. E., & Stathis, K. (2015). Recognising lifestyle activities of diabetic patients with a smartphone. In *9th International Conference on Pervasive Computing Technologies for Healthcare* (pp. 317–324).

Malawski, F., & Gałka, J. (2018). System for multimodal data acquisition for human action recognition. *Multimedia Tools and Applications*, *77*, 23825–23850.

Mimouna, A., Ben Khalifa, A., & BenAmara, N. E. (2018). Human action recognition using accelerometer data: Selective approach. In *International Multi-Conference on Systems, Signals & Devices* (pp. 467–472).

Namal, S., Senanayake, A., Vincent, C., James, C., & Rolland Sirisinghe, G. (2006). Analysis of soccer actions using wireless accelerometers. In *IEEE International Conference on Industrial Informatics, Singapore* (pp. 664–669).

Neto, P., Pires, J. N., & Moreira, A. P. (2009). Accelerometer-based control of an industrial robotic arm. In *The 18th IEEE International Symposium on Robot and Human Interactive Communication* (pp. 1192–1197).

Noor, M. H. M., Salcic, Z., & Wang, K. I.-K. (2015). Dynamic sliding window method for physical activity recognition using a single tri-axial accelerometer. In *IEEE 10th Conference on Industrial Electronics and Applications, ICIEA* (pp. 102–107).

Nuno, M., João, F., João, V., Mohammad, S., & Pedro, N. (2017). Human behavior and hand gesture classification for smart human-robot interaction. *Procedia Manufacturing*, *11*, 91–98.

Pepa, L., Verdini, F., Capecci, M., & Ceravolo, M. G. (2015). Smartphone based freezing of gait detection for Parkinsonian patients. In *IEEE International Conference on Consumer Electronics* (pp. 212–215).

Rastegari, A., Archenti, A,. & Mobin, M. (2017). Condition based maintenance of machine tools: Vibration monitoring of spindle units. In *Annual Reliability and Maintainability Symposium* (pp. 1–6).

Rodomagoulakis, I., Kardaris, N., Pitsikalis, V., Mavroudi, E., Katsamanis, A., Tsiami, A., & Maragos, P. (2016). Multimodal human action recognition in assistive human-robot interaction. In *IEEE International Conference on Acoustics, Speech and Signal Processing* (pp. 2702–2706).

Safi, K., Mohammed, S., Attal, F., Dedabrishvili, M., & Amirat, Y. (2016). Recognition of different daily living activities using hidden Markov model regression. In *3rd Middle East Conference on Biomedical Engineering* (pp. 16–19).

Sen, S., Vigneshwaran, S., Archan, M., Rajesh, K.B., & Youngki, L. (2015). The case for smartwatch-based diet monitoring. In *IEEE International Conference on Pervasive Computing and Communication Workshops* (pp. 585–590).

Tran, D. N., & Phan, D. D. (2016). Human activities recognition in android smartphone using support vector machine. In *7th International Conference on Intelligent Systems, Modelling and Simulation* (pp. 64–68).

Unuma, Y., & Komuro, T. (2015). Natural 3D interaction using a see-through mobile AR system. In *IEEE International Symposium on Mixed and Augmented Reality* (pp. 84–87).

Wang, Z., Donghui, W., Jianming, C., Ahmed, G., & Mohammad, A. H. (2016). A triaxial accelerometer-based human activity recognition via EEMD-based features and game-theory-based feature selection. *IEEE Sensors Journal*, *16*(9), 3198–3207.

Yunyoung Nam, Y. K. J. L. (2016). Sleep monitoring based on a tri-axial accelerometer and a pressure sensor. *Sensors*, *16*(5), 1–14.

Zhang, Y., Fei, Y., Xu, L., & Sun, G. (2015). Micro-IMU-based motion tracking system for virtual training. In *34th Chinese Control Conference* (pp. 7753–7758).

Zhang, L., Ou, M., Fu, X., & Yan, X. (2016). Using smartphones to estimate vehicle emission under urban traffic levels-of-service. In *12th World Congress on Intelligent Control and Automation* (pp. 1758–1763).

Zhao, X., Zhiming, G., Tao, F., Shishir, Shah, & Weidong, Shi. (2014). Continuous fine-grained arm action recognition using motion spectrum mixture models. *Electronics Letters*, *50*(22), 1633–1635.

Zia, A., Yachna, S., Vinay, B., Eric, L. S., & Irfan, E. (2018). Video and accelerometer-based motion analysis for automated surgical skills assessment. *International Journal of Computer Assisted Radiology and Surgery*, *13*(4), 443–455.

# Ultra Thin Nanocomposite In-Sole Pressure Sensor Matrix for Gait Analysis

**Dhivakar Rajendran, Bilel Ben Atitallah, Rajarajan Ramalingame, Roberto Bautista Quijano Jose, and Olfa Kanoun**

**Abstract** Gait analysis plays an important role in various applications such as health care, clinical rehabilitation, sport training and pedestrian navigation. In order to monitor the human gait, an interesting approach is to analyze the foot plantar pressure distribution between the foot and the ground. In recent years, the emergence of flexible, soft and lightweight sensors facilitates the rapid technological advances in in-shoe foot pressure measurements, thereby especially carbon nanotubes-based sensors provide an outstanding solution for the implementation of flexible, soft pressure sensors in foot pressure distribution analysis. This chapter focuses on the design and implementation of multiwalled carbon nanotubes (CNT)/polydimethylsil-oxane (PDMS) based nanocomposite pressure sensors for the analysis of the foot pressure distribution. The sensor is durable, stable and shows sensitivity of 3.3 kΩ/kPa and hysteresis smaller than 3.64% with maximum detectable pressure up to 217 kPa, which is suitable for the measurement of human foot pressure. The proposed sensor has been implemented in a flexible in-sole, which is designed based on normal arch foot anatomy. A total of 12 sensors are distributed in the heel, lateral back foot, mid-foot and front foot. The foot pressure distribution for different persons while walking and standing using nanocomposite sensor based in-sole were investigated by measuring the changing in resistance of the pressure sensors, when pressure applied on it. It shows that foot pressure distribution is higher in the fore foot and the heel while person standing in normal position. While walking, initially the foot pressure is in

D. Rajendran (✉) · B. Ben Atitallah · R. Ramalingame · R. B. Quijano Jose · O. Kanoun
Chemnitz University of Technology, Chemnitz, Germany
e-mail: dhivakar.rajendran@etit.tu-chemnitz.de

B. Ben Atitallah
e-mail: bilel.ben-atitallah@s2019.tu-chemnitz.de

R. Ramalingame
e-mail: rajarajan.ramalingame@etit.tu-chemnitz.de

R. B. Quijano Jose
e-mail: roberto.bautista@etit.tu-chemnitz.de

O. Kanoun
e-mail: olfa.kanoun@etit.tu-chemnitz.de

 
O. Kanoun and N. Derbel (eds.), *Advanced Sensors for Biomedical Applications*, Smart Sensors, Measurement and Instrumentation 38,
https://doi.org/10.1007/978-3-030-71225-9_2

the heel and then transferred to the entire foot and finally it is concentrated on the fore foot.

**Keywords** Carbon nanotubes · Polydimethylsiloxane · Nanocomposites · Pressure sensors · Screen printing · Solution mixing · Gait analysis

# 1 Introduction

The foot is important for humans for the interaction with ground for standing and locomotion. According to a institute of preventive foot health, an average human walks about 160,000 km, which is equal to 4 times around the earth (Franklin et al. 2015). In last decades, humans are subjected to various foot problems such as bunions, corns and calluses, hammertoes, heel pain, arch problems, chronic foot pain and diabetes related wounds (Crawford et al. 2020; Jeffcoate and Harding 2003). According to the International Diabetic Federation, 327 million people in age groups of 20–64 around the world and 58 million people in Europe were diagnosed for diabetes and it will be increased to 438 million people in 2045 (Federation 2020). Indeed, the diabetes mellitus accounts for over $1 billion per year in medical expenses (Federation 2020). Furthermore, gait instability in the elderly and other balance impaired individuals show the need of ways to analyze foot pressure distribution to obtain gait balance improvements which is considered important both in sports and biomedical applications such as forefoot loading during running, soccer balance training and foot balancing during weightlifting (Walther et al. 2020). It is estimated that around 13–59% people are subjected to foot injury during their daily activities (Walther et al. 2020).

In this chapter after a literature review on gait analysis, we introduce a low cost, highly durable, flexible binary nanocomposite consists of polydimethylsiloxane (PDMS) and multiwalled carbon nanotubes (MWCNT), which can reach high sensitivity and wide detectable pressure range. These sensors were implemented in a flexible in-sole where the design and sensor placement is based on the normal arch foot anatomy. This chapter concludes with evaluation of the in-sole by analysing the foot pressure distribution on a individual during walking and standing.

# 2 Gait Analysis

Gait analysis is important for providing the feet pressure distribution data to the physicians and therapists, to diagnose foot problems and walking disorders in order to find suitable treatments to patients and improving the gait stability during their sports and daily activities. To study the locomotion and kinematics of the foot, various models have been adopted to represent the gait pattern. In early stage, two dimensional models were considered to represent the foot. These models considered the foot as rigid body. Although the foot is a much more complicated structure with many individual muscle layers, bones, and joints (Lyons et al. 2006). During 1940s and 1950s, gait analysis was studied with the muscle activities during the different phase of the gait cycle by free-body diagrams and calculations on the effect of knee, hip and ankle joints (Lyons et al. 2006). In 1960s, many researchers employed mathematical modeling to demonstrate the motion of the body segments and actions of different muscles (Abdul Razak et al. 2012). Later, kinetograph, which is an apparatus for taking a series of photographs of moving objects for examination with the kinetoscope was introduced to measure the foot pressure distribution during walking motion in low cost measurement systems. The obtained kinetograph was analyzed by x-ray images and can be improved by placing a black rubber mat with reflecting pyramidal projecting fluid on a glass plate (Lyons et al. 2006). Further advances in gait analysis came with introduction of more accurate kinematic instruments in 1980s, which results in improved kinematic studies using electronics rather than images or visual observations that took a long time to gather information and also force platforms and EMED systems were made available, which produced reliable results in minutes (Zulkifli and Loh 2020). However, these advances in gait analysis are still expensive, as it requires dedicated equipment for the motion capture. Late 1990s, foot pressure analysis was enhanced using EMED sys-tem, F-scan systems, PEL-38 electronic podometer and piezoelectric sensors, which was placed under the foot and motion analysis camera has been used to investigate the gait pattern (Zulkifli and Loh 2020). However these systems were limited to the static foot pressure measurement.

In 21st century, various low-cost conventional techniques such as silicone layer, air bladder and optical fibers were used to enhance the foot pressure measurement in both static and dynamic mode (Kong and Tomizuka 2008; Soetanto et al. 2011). However, these techniques are not promising for the large deformation and repeating pressure cycle because of its limited mechanical strength. Later newer technology enabled electrical sensors to be implemented in the foot pressure measurement in two methods: platform and in-shoe. Platform systems are constructed from a flat, rigid array of pressure sensing elements arranged in a matrix configuration and embedded in the floor to allow and follow normal gait. Both static and dynamic measurement can be done in this system. But this system was restricted to the research laboratories and lack of on-field implementation (Abdul Razak et al. 2012). In-shoe sensors are flexible and embedded in the commercial shoes such that measurements reflect the interface between the foot and the shoe (Abdul Razak et al. 2012). Commercial sensors such as "Novel, Parotec, Tekscan, Vista Medical, Wahab" adopted both measurement systems to enhance the quality of the foot pressure measurement, where they can reach pressure range of 260 kPa–1.034 MPa and hysteresis of 0.05%–24% (Abdul Razak et al. 2012). The evolution of stretchable and flexible sensors paved the way for the many research works on the foot pressure analysis (Lou et al. 2017; Nobeshima et al. 2016; Pyo et al. 2017). Stretchable sensors based on conductive rubber, and conductive nanoparticles as nanofillers in soft polymers provides the flexible, durable and stable foot pressure measurement. These sensors have different layers, which act as artificial skin and it can reach sensitivity ranges from 0.1 W/N to 670 kW/N and detectable pressure from 3 to 600 kPa (Lou et al. 2017; Nobeshima et al. 2016; Pyo et al. 2017). To enhance the comfortability for the foot pressure measurement systems, these stretchable sensors are implemented in textiles "E-Textile" made of multiple layers such as carbon nano tubes coated polyester, graphene and organic polymers, which is suitable for the pressure range 0.01–1.2 MPa (Lin et al. 2016; Lou et al. 2017). Despite its comfort, it is not durable for long time loading application. In most of the works based on stretchable sensors employing MWCNT as nanofillers, one of the best choice for matrix are soft polymers Thermoplastic poly urethane, polyester, polyethylene terephthalate and PDMS (Polydimethylsiloxane). This is due to that the soft polymer has low Young's modulus can retain to its orginal form faster than other polymers and it can be utilized for dynamic and high frequency foot pressure applications like walking and running (Canavese et al. 2014; Cheng et al. 2011; da Costa and Choi 2017; Huang et al. 2017; Karimov et al. 2015; Lee and Choi 2008; Ramalingame et al. 2017a; Sepulveda et al. 2011; Shu et al. 2010; So et al. 2013).

## 3 CNT/Polymer Pressure Sensor

Recently, various research works on CNT-based sensors, have been highly active in interest for diverse applications. Such composites are obtained by introducing enough dispersed CNTs into a polymer matrix that enables sensing capabilities in the resulting nanocomposite. The conductivity and thus, the sensing properties of these composites depend on numerous factors, such as the quality and kind of the polymer matrix and the size, nature and concentration of the dispersed CNTs. When compressive forces are transferred to the surface of these nanocomposites, the distributed conductive particles are induced to contact each other, resulting in formation of more conducting paths than the already existing before applying the pressure and hence reducing the electrical resistivity of the nanocomposite (Fig. 1).

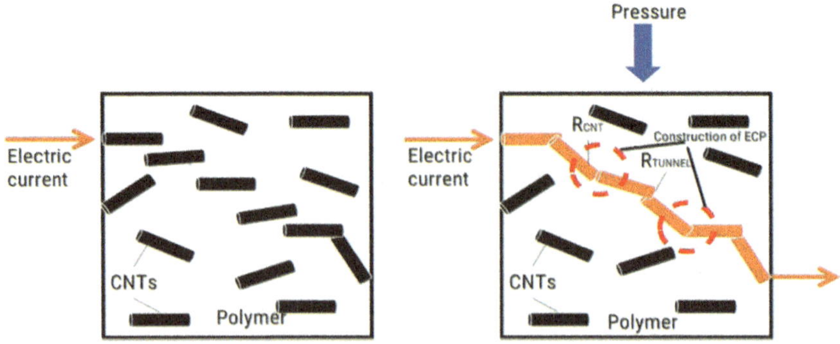


**Fig. 1** Resistve principle of CNT/polymer nanocomposite pressure sensor

CNTs have been already incorporated in various polymer such as poly (methyl methacrylate) (PMMA), polycarbonate (PC), poly(L-lactide) (PLLA), pol vinyl alcohol (PVA) and others, in order to fabricate pressure sensors (Sousa et al. 2015). Among other nanocomposites, medical grade polydimethylsiloxane (PDMS) added with CNTs results in a nanocomposite material with high biocompatibility and flexibility, which suits pressure sensing application involving direct human contact (Maddipatla et al. 2017).

## 3.1 Fabrication Process of CNT/PDMS Pressure Sensors

Solution mixing approach is used to fabricate the CNT/PDMS nanocomposites, which involves sonication and mechanical stirring. These methods provide adequate stress to de-bundle the CNTs and disperse it in the PDMS effectively (Ramalingame et al. 2019). Tetrahydrofurane (THF) is used as common organic solvents, because it is one of the optimum solvent to disperse the CNTs efficiently (Ramalingame et al. 2017b). 0.3 wt. % MWCNT (Sigma-Aldrich, O. D$\times$L—4.5 nm $\times$0.5 nm $\pm$3–6 $\mu$m, 95%) with THF and sonicated using Bandelin sonoplus for 20% of the total power for 15 min and then stirred magnetically using CAT-M27 for 60 min, then mixed with PDMS (CNT/THF:PDMS—1:1).

The mixture is sonicated for 30 min at 50% amplitude and then stirred magnetically for 60 min at 70 °C followed by sonication of 50% amplitude for 30 min (Fig. 2).

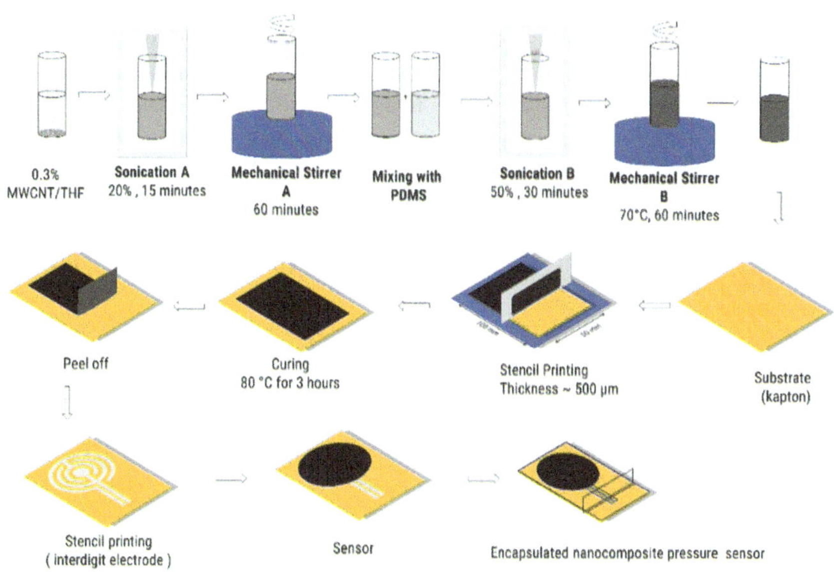


**Fig. 2** Fabrication of CNT/PDMS nanocomposite pressure sensor

Curing agent is added at the ratio of 10:1 and mixed manually for 15 min to mix hardener uniformly in nanocomposite. Then, thin film is deposited over the Kapton substrate using stencils printing with thickness of 200 μm in the rate 1 mm/s. Then it is allowing for curing for 80 °C for 4 h. Once it is cured its peeled off from the Kapton and diced in 10 mm radius. In parallel, silver based interdigit electrode is printed over Kapton using stencil printing at the rate of 1 mm/s and cured at room temperature for 4 h. Then thin film is placed over the electrode and encapsulated using PET with the help of commercially available laminator.

The electrical conductivity of the MWCNT/PDMS nanocomposite pressure sensor is measured using a digital multimeter Agilent 24401A. The conduction mechanism is based on the percolation theory and tunnelling effect where CNTs act like metal-metal junctions form conduction paths between the electrode and polymer will act like tunnelling barrier between CNTs. This phenomenon can be visualized as three-dimensional resistors networks based on tunnelling effect (Kanoun et al. 2021). Theoretically, two types of resistance can be seen in MWCNTs-PDMS nanocomposite films such as $R_{tube}$ is intrinsic resistance of CNT, which will be around 0.2 kΩ/μm to 0.4 kΩ/μm (Kanoun et al. 2021). The second type is Rjunction, which is further divided into contact resistance between CNTs ($R_c$) and tunnelling resistance ($R_T$) between CNTs, which can be seen below:

$$R = R_{tube} + R_{junction} \tag{1}$$

$$R_T = \frac{h^2 d}{Ae^2\sqrt{2m\lambda}} e^{\frac{4\pi d}{h}\sqrt{2m\lambda}} \tag{2}$$

where $d$ is the distance between CNT, $h$ is the Planck constant, $e$ is the quantum of electricity, $l$ is the barrier height of energy, $m$ is the electron mass and $A$ is the cross-sectional area of the tunnel (Sepulveda et al. 2011). During initial applied force of 1N, the nanocomposite sensor comes in contact with the underlaying electrode, which results in abrupt decrease in resistance of the sensor. So, the sensor needs an activation force of 1N to provide a stable measurement. Further increases in force results in further decreases in resistance.

This behaviour can be explained by Eqs. (1) and (2), where the change in orientation of conduction path results in change of $R_c$, change in tunnelling distance results in change of $R_T$, which in turn change the $R_{junction}$ and deformation of CNTs change in $R_{tube}$, which results in overall resistance decreases (Kanoun et al. 2021). From Fig. 3 it can be seen that the change in resistance gradually reduced at high force ($< \approx 100$ kPa), which is due to less formation of conductive paths and once it reaches a saturation point, there is no formation of new conductive paths (Kanoun et al. 2014). At this point, the change in resistance at high pressure region is due to the intrinsic resistance of CNTs, being slightly compressed.

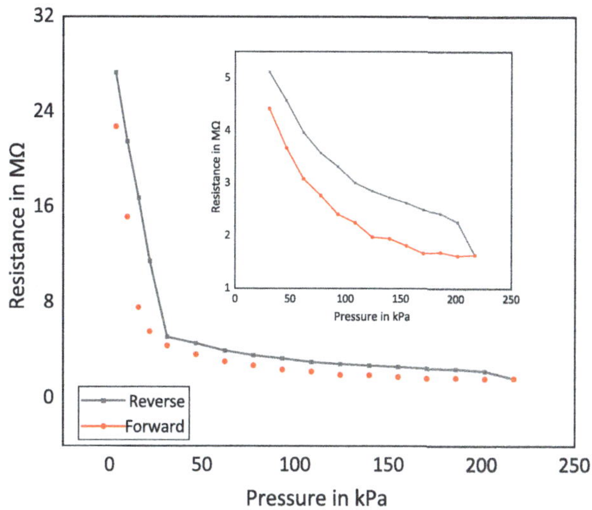


**Fig. 3** Sensor behaviour for forward and reverse force cycle. Inside: Sensor behaviour for force of 3–217 kPa

# 4 Implementation and Evaluation

In-sole design and sensor placement are mainly dependent on foot anatomy like flat or high arched foot, clubfoot and an extra toe. Normally, a person is allowed to walk over the PressureStat film during a normal stride where these films will give the exact replica of the person footprint and sensors are placed in maximum pressure bearing points, which is reflected as darken area in Pressurestat film (Stalin 2012). In the work reported here, in-sole design is based on normal arch foot anatomy (UK size 10) and sensor is placed based on the pressure distribution for that foot taken from foot anatomy. A total of 12 sensors are distributed in the heel, lateral back foot, midfoot and foot. In heel, three sensors are placed. Pressure distribution is less in mid foot and so one sensor is placed. During the gait, the last phase of 1st stance will be on the forefoot and so seven sensors are placed in this region to get more resolution of pressure distribution in this region. The electrode layout is designed in adobe illustrator 2020. Then the electrodes are fabricated by silver inkjet printing using Diamtrix DMP-2850 obtaining 12 interdigitated electrodes plus a common ground. Next the nanocomposite films were prepared using fabrication technique explained above are placed over the electrodes and finally encapsulated with PET. Figure 4 shows the electrodes layout and sensors placement on the in-sole; the sensors were numbered as $S_1$ to $S_{12}$ according to their location in the in-sole.

To test and validate our system, we first implement a measuring system with the help of Arduino microcontroller and a voltage divider circuit. A young adult subject participated in the controlled experiment to validate the system. Pressure distribution on the foot were analysed in stationary and dynamic phase. During stationary phase,

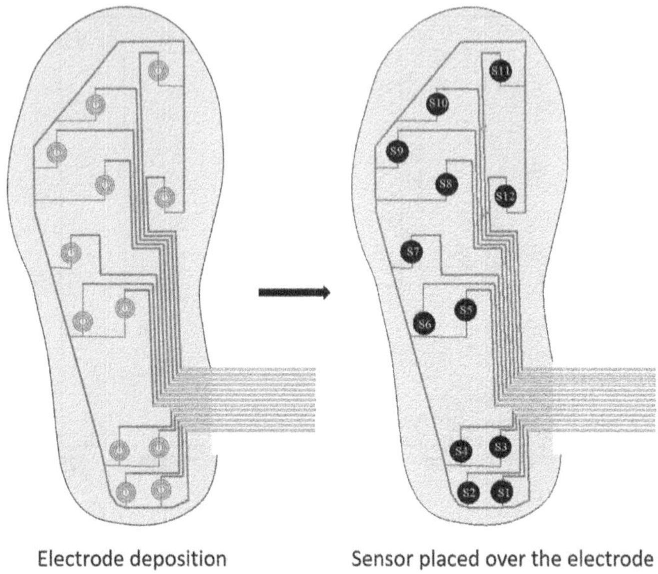


**Fig. 4** Electrodes layout and sensors placement on the in-sole

the in-sole was placed on the ground and the subject could stand on it. During the dynamic phase, the insole was placed in each of the subject's shoes, and the subject was asked to perform different tests. Foot pressure data was recorded at time interval of 50 ms in the excel file and then extracted from the database after the experiment for further analysis.

## *4.1 During Stationary Phase*

The aim of this experiment was to investigate the pressure distribution of the different sections of the foot such as forefoot, mid foot and heel at the stationary scenario. The subject was asked to do this test of each position for 7 s and 5 s gap in-between the different positions to ensure reliable and accurate results.

Figure 5 shows the maximum change in relative resistance of the pressure sensors placed on the in-sole when the person is asked to stand on it in different stationary position such as initial contact, loading response and heel off, all the positions are performed on the other leg support. Higher change in relative resistance means that more pressure is applied on the respective position. In initial contact position, the change in relative resistance of the sensors in the heel position are −79% to −86%. This is because in initial contact position, only the heel is in contact with the ground (in-sole) and most of the pressure will be applied on the heel position.

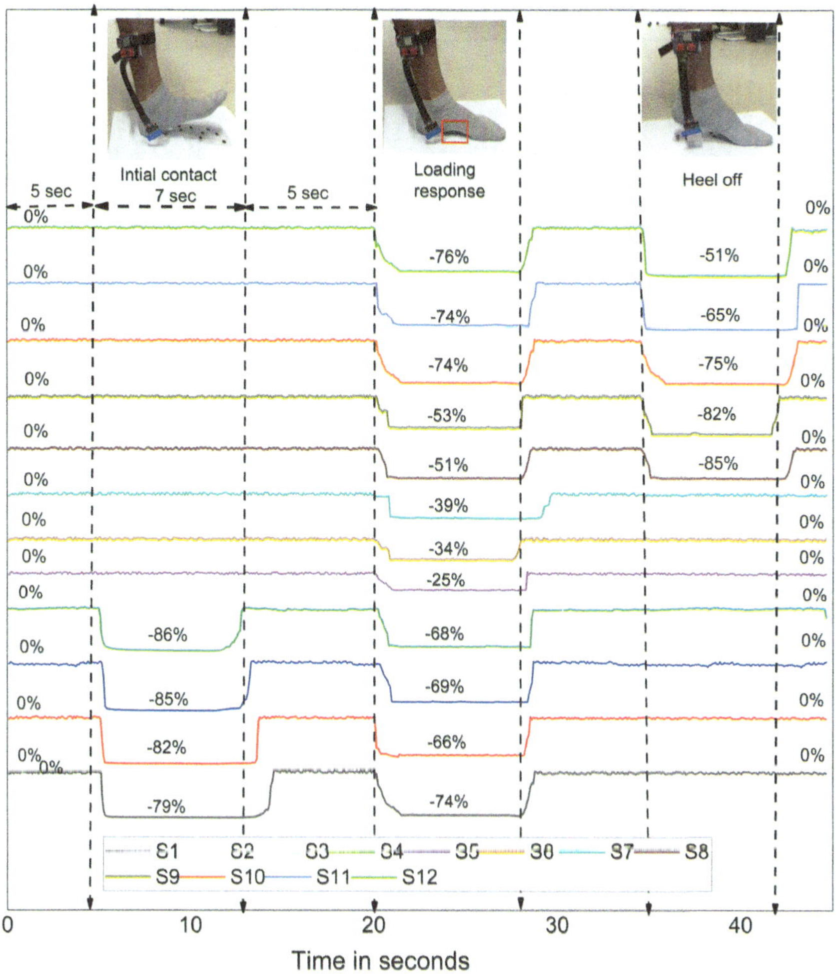


**Fig. 5** Varying foot level like walking in stationary phase

## *4.2 During Dynamic Phase*

The aim of this experiment was to investigate the dynamic gait pattern on a single leg during normal walking, which is considered as swinging leg and other one is the supporting leg. The gait pattern is comprised by two stances, each stance has five steps such as initial contact, load response, mid distance, terminal distance and pre-swing. In this work, the signal is recorded for the 1st stance of the walking (Fig. 6).

Figure 7 shows that maximum change in relative resistance of pressure sensors placed on the in-sole when the person subjected to normal walking. The first step of

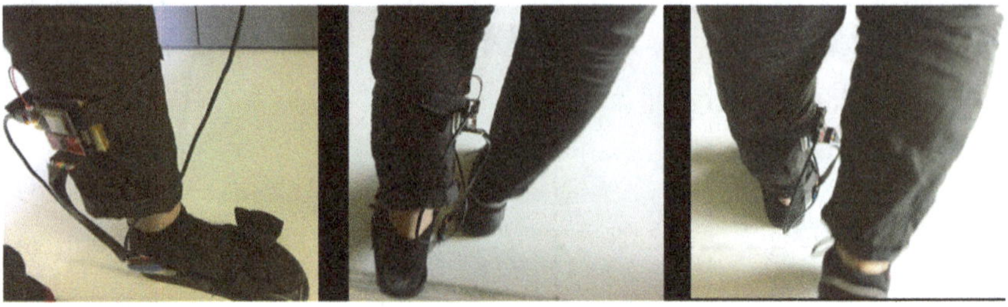

Fig. 6 Evaluation of electronic in-sole in dynamic phase

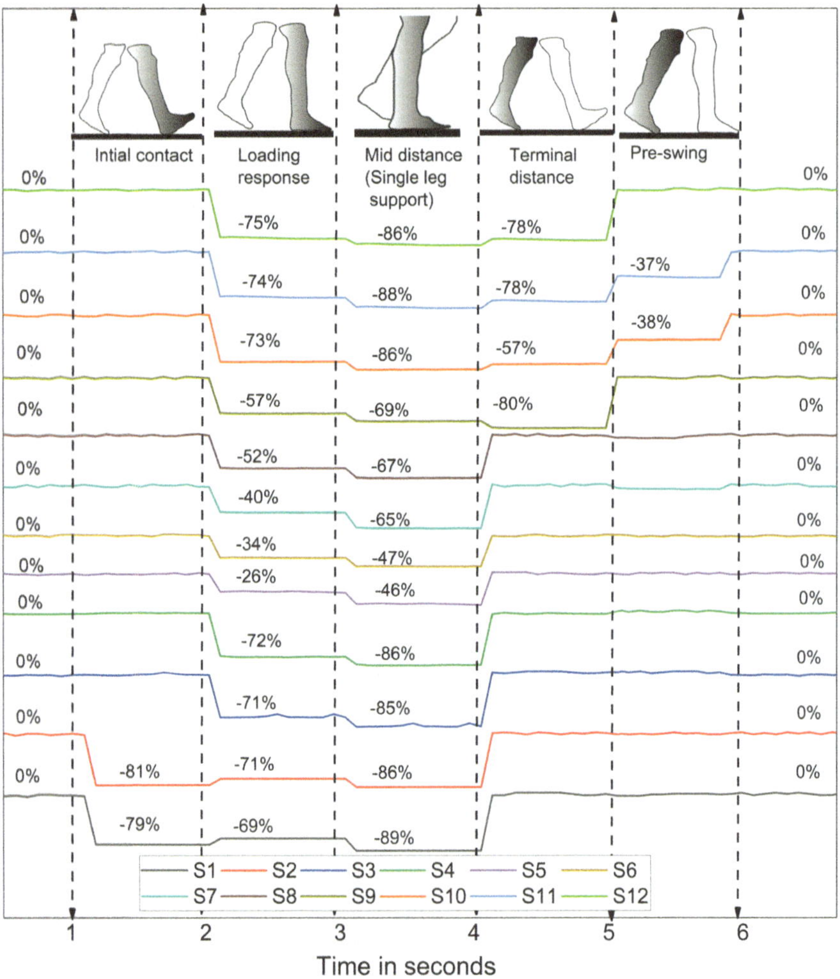


Fig. 7 Foot pressure distribution in different dynamic phase

1st instance of normal walking is initial contact where the body weight is completely focused on the heel of the swinging leg and so more pressure is applied on it. The next instance is when the body weight is transferred and distributed from the heel to the mid and fore foot of swinging leg. Despite of the pressure distribution, more pressure is applied on the fore foot and the heel. When the swinging leg reaches mid distance position, the respective foot is subjected to more pressure because the body weight is transferred from the supporting foot while it is not in contact with the ground. This can be seen from Fig. 7 that the change in relative resistance of the swinging foot is increased from in the range of 10–15% from the previous position. In terminal distance, the heel of the supporting foot and fore foot of the swinging foot will come to contact with the ground.

The body weight is transferred on both feet and so the pressure applied on the fore foot of the swinging leg is reduced compared to the mid distance position. This can be seen as, the change in relative resistance of the swinging leg is re-duced further in the range of 35%–45%.

## 5 Conclusion

In this chapter, after a brief analysis on various methods for monitoring the gait pattern and foot pressure distribution, we introduced pressure sensor based on Multiwalled carbon nanotubes (MWCNT) and polydimethylsiloxane (PDMS) with good sensitivity and wide detectable pressure range for gait analysis. We reported the design and fabrication technique of nanocomposites and pressure sensor based on CNT polymer nanocomposites in order to achieve a low cost, highly durable and highly sensitive sensing in-sole. The proposed sensors show sensitivity of $3.3\,\mathrm{k}\Omega/\mathrm{kPa}$ in the range of 110 kPa and hysteresis of <3.64% with maximum detectable pressure up to 217 kPa. This chapter concluded with analysis of the foot pressure distribution while the person standing in stationary phase and gait pattern of the person during walking.

**Acknowledgements** The research work was carried out under the project "SenseCare—high-tech sensor technology for the challenge of demographic change in Saxony (100270070)" funded by the European Social Fund and the Free State of Saxony, "Landesinnovationsstipendium (100284169)", funded by the Sächsische Aufbaubank (SAB) and the European Social Fund (ESF) and by the Deutsche Forschungsgemeinschaft (DFG, German Research Foundation)—Project-ID 416228727—SFB 1410, applicant Prof. Dr.-Ing. Olfa Kanoun.

Europäische Union

Europa fördert Sachsen.
ESF
Europäischer Sozialfonds

## References

Abdul Razak, A. H., Zayegh, A., Begg, R. K., & Wahab, Y. (2012). Foot plantar pressure measurement system: A review. *Sensors, 12*(7), 9884–9912.

Canavese, G., Stassi, S., Fallauto, C., Corbellini, S., Cauda, V., Di Donato, M., Pirola, M., & Pirri, F. C. (2014). Stretchable and wearable piezoresistive insole for continuous pressure monitoring. *Key Engineering Materials, 605*, 474–477. Trans Tech Publications

Cheng, M.-Y., Tsao, C.-M., Lai, Y.-Z., & Yang, Y.-J. (2011). The development of a highly twistable tactile sensing array with stretchable helical electrodes. *Sensors and Actuators A: Physical, 166*(2), 226–233.

Crawford, F., Nicolson, D. J., Amanna, A. E., Martin, A., Gupta, S., Leese, G. P., et al. (2020). Preventing foot ulceration in diabetes: Systematic review and meta-analyses of rct data. *Diabetologia, 63*(1), 49–64.

da Costa, T. H., & Choi, J.-W. (2017). A flexible two dimensional force sensor using PDMS nanocomposite. *Microelectronic Engineering, 174*, 64–69.

Federation, I. D. (2020). *IDF Diabetes Atlas, 9th edn.*, Brussels, Belgium. https://www.diabetesatlas.org. Accessed 2 July 2019.

Franklin, S., Grey, M. J., Heneghan, N., Bowen, L., & Li, F.-X. (2015). Barefoot vs common footwear: A systematic review of the kinematic, kinetic and muscle activity differences during walking. *Gait & Posture, 42*(3), 230–239.

Huang, W., Dai, K., Zhai, Y., Liu, H., Zhan, P., Gao, J., et al. (2017). Flexible and lightweight pressure sensor based on carbon nanotube/thermoplastic polyurethane-aligned conductive foam with superior compressibility and stability. *ACS Applied Materials & Interfaces, 9*(48), 42266–42277.

Jeffcoate, W. J., & Harding, K. G. (2003). Diabetic foot ulcers. *The Lancet, 361*(9368), 1545–1551.

Kanoun, O., Bouhamed, A., Ramalingame, R., Bautista-Quijano, J. R., Rajendran, D., & Al-Hamry, A. (2021). Review on conductive polymer/CNTs nanocomposites based flexible and stretchable strain and pressure sensors. *Sensors, 21*(2), 341; 1–29.

Kanoun, O., Müller, C., Benchirouf, A., Sanli, A., Dinh, T. N., Al-Hamry, A., et al. (2014). Flexible carbon nanotube films for high performance strain sensors. *Sensors, 14*(6), 10042–10071.

Karimov, K. S., Sulaiman, K., Ahmad, Z., Akhmedov, K., & Mateen, A. (2015). Novel pressure and displacement sensors based on carbon nanotubes. *Chinese Physics B, 24*(1)

Kong, K., & Tomizuka, M. (2008). Smooth and continuous human gait phase detection based on foot pressure patterns. In *2008 IEEE International Conference on Robotics and Automation* (pp. 3678–3683). IEEE.

Lee, D.-W., & Choi, Y.-S. (2008). A novel pressure sensor with a PDMS diaphragm. *Microelectronic Engineering, 85*(5–6), 1054–1058.

Lin, F., Wang, A., Zhuang, Y., Tomita, M. R., & Xu, W. (2016). Smart insole: A wearable sensor device for unobtrusive gait monitoring in daily life. *IEEE Transactions on Industrial Informatics, 12*(6), 2281–2291.

Lou, C., Wang, S., Liang, T., Pang, C., Huang, L., Run, M., et al. (2017). A graphene-based flexible pressure sensor with applications to plantar pressure measurement and gait analysis. *Materials, 10*(9), 1068.

Lyons, T. E., Rosenblum, B. I., & Veves, A. (2006). Foot pressure abnormalities in the diabetic foot. In *The Diabetic Foot*, (pp. 163–184). Berlin: Springer.

Maddipatla, D., Narakathu, B. B., Ali, M. M., Chlaihawi, A. A., & Atashbar, M. Z. (2017). Development of a novel carbon nanotube based printed and flexible pressure sensor. In *2017 IEEE Sensors Applications Symposium (SAS)* (pp. 1–4). IEEE.

Nobeshima, T., Uemura, S., Yoshida, M., & Kamata, T. (2016). Stretchable conductor from oriented short conductive fibers for wiring soft electronics. *Polymer Bulletin, 73*(9), 2521–2529.

Pyo, S., Jo, E., Kwon, D.-S., Kim, W., Chang, W., & Kim, J. (2017). Fabrication of carbon nanotube-coated fabric for highly sensitive pressure sensor. In *2017 19th International Conference on Solid-State Sensors, Actuators and Microsystems (TRANSDUCERS)* (pp. 962–965). IEEE.

Ramalingame, R., Hu, Z., Gerlach, C., & Kanoun, O. (2017a). Shoe insole with mwcnt-pdms-composite sensors for pressure monitoring. In *2017 IEEE Sensors* (pp. 1–3). IEEE.

Ramalingame, R., Rajendran, D., & Kanoun, O. (2017b). Method optimization of mwnct/pdms nanocomposites using organic solvents. Printed Future Days 2017 (pp. 121–125).

Ramalingame, R., Lakshmanan, A., Müller, F., Thomas, U., & Kanoun, O. (2019). Highly sensitive capacitive pressure sensors for robotic applications based on carbon nanotubes and pdms polymer nanocomposite. *Journal of Sensors and Sensor Systems*, *8*(1), 87–94.

Sepulveda, A. T., Fachin, F., de Villoria, R. G., Wardle, B. L., Viana, J. C., Pontes, A. J., et al. (2011). Nanocomposite flexible pressure sensor for biomedical applications. *Procedia Engineering*, *25*, 140–143.

Shu, L., Hua, T., Wang, Y., Li, Q., Feng, D. D., & Tao, X. (2010). In-shoe plantar pressure measurement and analysis system based on fabric pressure sensing array. *IEEE Transactions on Information Technology in Biomedicine*, *14*(3), 767–775.

Soetanto, W., Nguyen, N. T., & Wang, W.-C. (2011). Fiber optic plantar pressure, shear sensor. In *Health Monitoring of Structural and Biological Systems 2011* (Vol. 7984, p. 79840Z). International society for optics and photonics

So, H.-M., Sim, J. W., Kwon, J., Yun, J., Baik, S., & Chang, W. S. (2013). Carbon nanotube based pressure sensor for flexible electronics. *Materials Research Bulletin*, *48*(12), 5036–5039.

Sousa, P. J., Silva, L. R., Goncalves, L. M., & Minas, G. (2015). Patterned CNT-PDMS nanocomposites for flexible pressure sensors. In *2015 IEEE 4th Portuguese Meeting on Bioengineering (ENBENG)* (pp. 1–4). IEEE.

Stalin, M. (2012). *Development of a smart insole system for real-time detection of temporal gait parameters and related deviations in unilateral lower-limb amputees*. Doctoral Dissertatin, University of Miami.

Walther, M., Hörterer, H., & Hilgers, M. (2020). Foot injuries. In *Injury and health risk management in sports* (pp. 173–178). Berlin: Springer.

Zulkifli, S. S., & Loh, W. P. (2020). A state-of-the-art review of foot pressure. *Foot and Ankle Surgery*, *26*(1), 25–32.

# Piezo-Resistive Pressure and Strain Sensors for Biomedical and Tele-Manipulation Applications

**Bilel Ben Atitallah, Dhivakar Rajendran, Zheng Hu, Rajarajan Ramalingame, Roberto Bautista Quijano Jose, Renato da Veiga Torres, Dhouha Bouchaala, Nabil Derbel, and Olfa Kanoun**

**Abstract** Nowadays, the demand for flexible and wearable devices is significantly increasing. Thereby, sensors based on carbon materials are gaining importance due to their high flexibility, sensitivity and medical compatibility. Precisely, nanocomposite based pressure and strain sensors present an interesting potential of applied force detection that helps to build the basis for body attached sensor networks. These sensor principles based on polymer carbon nanotubes composites (PCN) will have the capability for tracking finger movements, gestures and grasping. Therefore, several studies are explored in hand muscle rehabilitation, sign communication and

B. Ben Atitallah (✉) · D. Rajendran · Z. Hu · R. Ramalingame · R. B. Quijano Jose · R. da Veiga Torres · O. Kanoun
Professur Mess-und Sensortechnik, Technische Universität Chemnitz, Chemnitz, Germany
e-mail: bilel.ben-atitallah@s2019.tu-chemnitz.de

D. Rajendran
e-mail: dhivakar.rajendran@etit.tu-chemnitz.de

Z. Hu
e-mail: zheng.hu@etit.tu-chemnitz.de

R. Ramalingame
e-mail: rajarajan.ramalingame@etit.tu-chemnitz.de

R. B. Quijano Jose
e-mail: roberto.bautista@etit.tu-chemnitz.de

R. da Veiga Torres
e-mail: renato-da-veiga.torres@s2015.tu-chemnitz.de

O. Kanoun
e-mail: olfa.kanoun@etit.tu-chemnitz.de

D. Bouchaala
Digital Research Center of Sfax, University of Sfax, Sfax, Tunisia
e-mail: bouchaala.dhouha@gmail.com

N. Derbel
CEM Research Laboratory, National School of Engineer of Sfax, University of Sfax, Sfax, Tunisia
e-mail: n.derbel@enis.run.tn



O. Kanoun and N. Derbel (eds.), *Advanced Sensors for Biomedical Applications*, Smart Sensors, Measurement and Instrumentation 38,
https://doi.org/10.1007/978-3-030-71225-9_3

robotic telemanipulation. This chapter reviews developed carbon materials sensors and integrated solutions for hand gestures/forces detection in the biomedical application and robotic telemanipulation. In this context, a novel PCN strain and pressure sensors were presented and investigated. An SBS (styrene-butadiene-styrene rubber)/C-TPU (conductive thermoplastic urethane) strain sensor with 1 mm as the diameter is developed. The proposed sensor shows promising sensitivity and stretchability performances with up to 50% of strain and gauge factor equal to 24. In the other part, Poly-Dimethylsiloxane (PDMS)/Multiwalled carbon nanotubes (MWCNTs) pressure sensors are investigated. The results demonstrate excellent sensing performance e.g. fast response to detect low pressure, high durability after 100 cyclic loading/unloading test and high sensitivity up to 670 kPa. Moreover, a hybrid hand motion detection system was implemented for hand rehabilitation and gestures detection. The proposed sensors were attached to a glove that leads to the monitoring of fingers' movements and the palm pressure distribution.

**Keywords** Nanocomposite–sensors · Strain · Pressure · Rehabilitation · Tele-manipulation · Bio-application · Piezo-resistive · CNT · Hand gloves

## 1 Introduction

The demands of flexible, easy to use and energy-efficient electronic devices as well as up to date technologies in health care monitoring are increasing in medical investigation and patient's therapy (Williams et al. 2016). A statistics shows that there are a huge numbers among existing people in this planet suffering from the muscle diseases such us Williams syndrome (estimated to be affects 1 in 10,000 people worldwide Lenhoff et al. 1997), and Stroke (about four million people are surviving from stroke, and it is increasing reaching approximately 500,000 people each year Boian et al. 2002) etc. Medical investigations studies demonstrate that mostly of these people suffer from significant disabilities in their motor capabilities, which limits the patient's independent living. The "Bureau of Labor Statistics" reported for stork survivors over 143,000 hand-related workplace injuries happened in 2015, and it is the second most common body part to be injured (Burtney 2020). Despite increasing hand-related problems around the world, only 3% of total health care systems is devoted to the rehabilitation process (Boian et al. 2002). In order to recover partially or totally the motor abilities of a patient, a rehabilitation process is required which can take a few months to improve the ability of moving the hand in a normal way(Mulas et al. 2005). Patients need intensive and repetitive tasks during their rehabilitation until their hands reach the highest level of independence. Research has revealed that duration, capacity, and intensity of the training session have a significant impact on the motor rehabilitation improvement. In recent years, a deep research on rehabilitation engineering applications was announced to provide a cost-efficient, comfort and user-friendly rehabilitation systems, aims to enhance the quality of the rehabilitation process. Stilli et al. (2018) develop a novel light-weight inflatable soft exoskeleton device for post-stroke patients. A pneumatic solution is proposed with comprehensive mathematical model for hand sizes characterisation, which offers high-dosage, adaptive and gradual rehabilitation therapy. Park et al. (2018) proposed a multimodal sensing and interaction paradigm for a hand orthesis incorporating EMG armband, bend and pressure commercial sensors. EMG is complemented by others sensing modalities, where the bend and pressure provide the system information about proposed rehabilitation movements. In this proof-of-concept implementation, several diseases phenomena can be recorded such as muscle spasticity and abnormal muscle synergies, which can enhance the medical understanding. Ergen and Oksuz (2019) performed a grip analyses in order to investigate the load distribution and the contact area and grasped objects. A special glove based attached pressure sensors was developed for collecting and recording instant data. This study was announced in hand assessment area, where four common grip types (standard, lateral, pinch, and tripod grips) were investigated in terms of load distribution and contact area related to the palmar surface of the hand. Karime et al. (2011) develop an interactive rehabilitation system based on an IMU sensor that is aimed to help in the recovery process of people with wrist disabilities. A 2D golf game was designed as exercises platform, which leads patient to perform daily exercises in joyful and interactive manners. Dovat et al. (2008) developed a robotic interface based on biomechanical measurements, which can assist the subject in opening and closing movements and can be adapted to accommodate various hand shapes and finger sizes.

## 2 Nanocomposite Sensors

Among a number of smart nano materials, carbon nanotubes (CNTs) present an important role in force sensing area due to their remarkable electrical, mechanical, and electrochemical properties. Carbon nanotubes were classified to be outstanding fillers materials in high-performance polymer composites for interaction with the human body (Kanoun et al. 2021). In order to understand the conductive mechanism of the nanocomposite pressure/strain sensors, the change of the CNT/polymer interaction network under applied forces have to be studied first. In fact, the CNT nano-fillers are usually dispersed in a polymer to improve the nanocomposite properties. The carbon nanotubes material is a rigid material with high electrical and thermal properties. In general, a non-conductive polymer will be mixed to enhance mechanical and dielectric properties. In this context, Dinh et al. (2011) proved that the sensitivity of the carbon nanotubes embedded in the polymer is higher than the sensitivity of the CNT films without embedded/integrated polymer. The nanocomposite sensors are built up of different types of polymer, which differs based on the measurement range and the physical quantity measured.

## 2.1 Pressure Sensors

### 2.1.1 Sensor Principle

After the nanocomposite fabrication, not all CNTs will participate in the conductive network formation due to the random dispersion in the polymer matrix (see Fig. 1). The nanocomposite conductivity is induced by the CNT network by establishing conducting paths between electrodes. Based on the two-dimensional stick percolation theory, each CNT is defined as straight stick responsible in constructing the conductive network of the composite. With the increase of the conducting filler content, the nanocomposite material pass from the insulation state to a conductive state (Kanoun et al. 2014). This enables to accelerate the formation of the conductive network, which significantly, increases the magnitude of the current carrying that passes through the conductive network (see Fig. 1). When the sensor is under pressure, position, and orientation of the carbon nanotubes in the resistance network change to cause a significant change of the conductive paths. Also, the applied pressure creates a modification in both the geometry and the area of the CNT film. All these factors cause a resistance variation of the nanocomposite pressure sensor.

As illustrated in Fig. 1, CNTs components were randomly dispersed in the polymer network. These CNTs are connected by tunnelling transport of electron via junction gaps (Wang and Ye 2013). Two types of resistance are defined to estimate the resistance of a CNT film as expressed in the following:

$$R_{Total} = R_{CNT} + R_{junction} \tag{1}$$

where $R_{CNT}$ is the intrinsic resistance of the carbon nanotubes itself, and $R_{junction}$ is the inter-tube resistance resultant from the CNT/polymer interaction.

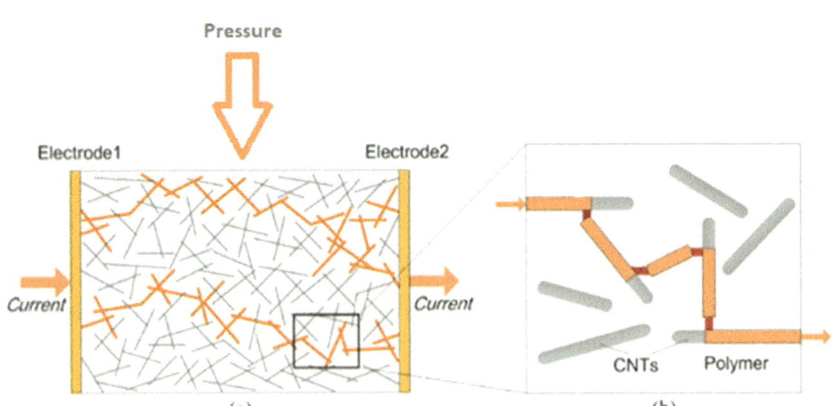


**Fig. 1** Electrical conductive CNT network in a polymer matrix under pressure

The typical value for $R_{CNT}$ is estimated between 0.2 kΩ/μm and 0.4 kΩ/μm. The $R_{junction}$ resistance can be divided into two principle resistance $R_{contact}$ and $R_T$, where $R_{contact}$ is the physical contact resistance with the CNTs as calculated in Eq. (2):

$$R_{contact} = R_{ij} = \frac{4l_{ij}}{g_{CNT}\pi d} \quad (2)$$

The $R_T$ is the tunneling resistance for carbon nanotubes separated by small gaps as expressed in Eq. 3:

$$R_T = R_{ij} = \frac{h^2 d}{Ae^2\sqrt{2m\lambda}} e^{\frac{4\pi d}{h}\sqrt{2m\lambda}} \quad (3)$$

where $d$ is the distance between CNT, $h$ is the plank constant, $e$ is the electricity quantum, $m$ is the electron mass, $\lambda$ is the energy barrier and $A$ is the cross sectional area of the tunnel.

### 2.1.2 Recent Developments of Pressure Sensors: An Overview

Dinh et al. (2018) fabricated a pressure sensor based on carbon nanotubes yarns by dry web-spinning and heat-treatment processes. A chemical vapor deposition (CVD) process was used to achieve the growth of CNT forests on the silicon substrate. The CNT forests spurned into yarns by a dry spinning process to results CNT yarns with 12 μm of diameter. Then, to reduce the van der Waals interaction between adjacent CNT, the CNT web on the silicon substrate was annealed in a furnace where the temperature varies between 200 °C and 600 °C. The CNT yarns were mixed into a web of acrylic elastomer using a thickness of 500 μm. Experimental results exhibit a high electrical and mechanical performance, where the resistance decrease by 2.8% at a pressure threshold lower than 60 kPa, a high sensitivity detected and Young's modulus was estimated approximately on 4.7 $10^{-4}$ $kPa^{-1}$ and 16 GPa.

Jung et al. (2018) proposed a new concept of a nanocomposite pressure sensor where a pyramidal protruding CNT/PDMS structure was developed. The nanocomposite material is restricted between two conductive electrodes due to the great electrical properties of PDMS polymer. The top electrode and the bottom electrode are fabricated by the polyethylene terephthalate (PET) films, which coated with tin-indium oxide (ITO) by the means of the high conductivity characteristic. During the unloading period, only the tip of the pyramid was in contact with the bottom electrode. Contrariwise, the surface of contact increases under pressure, accordingly, the resistance measurement decrease due to the augmentation of the CNT conductive paths in the resistor network. Experimental results demonstrate three different linear sections according to the pressure applied. When the quantity of pressure subjected between 0 to 25 Pa, the sensor exhibits a sensitivity of 21,14 $kPa^{-1}$, between 25 Pa to 250 Pa of pressure the sensitivity decrease to 1.39 $kPa^{-1}$ and a low sensitivity detected when applied more than 250 Pa approximated to 0.08 $kPa^{-1}$. Gao et al. (2018) designed a wearable pressure sensor based on sandpaper-moulded carbon nanotube-polydimethylsiloxane (CNT-PDMS). Using the sandpaper moulding method owing to its high efficiency and scalability, the pressure sensor was developed. First, the PDMS polymer was deposed onto the sandpaper with a grit number of 1200. The microstructures are called to the PDMS layer and peeled off from the sandpaper. Then the electrode was finished by coating the CNTs onto the sandpaper moulded PDMS. Two proposed electrodes were needed and placed face to face to form the final state of the pressure sensor. The device was tested under pressure applied between 5 Pa and 50 kPa. The nanocomposite sensor illustrates a response time approximated to 190 ms, a sensitivity of 0.2 $kPa^{-1}$ at the limit pressure and high stability where more than 5000 loading/unloading cycle are done.

Liu and Yan (2018) demonstrated a successful fabrication of a nanocomposite pressure sensor based on the capacitance variation. The capillary pressure sensor was composed of two parallel plates based on CNT material and PDMS polymer including a medium liquid as a dielectric medium. Where the capacitance was varied due to the variation of the interspace between electrodes under pressure. The fabrication process requires 3 g of MWCNT with a diameter between 5–8 nm and length between 10–30 μm were dispersed in 50 g of ethanol using a mechanical homogenizer. After defined time, a 32 g of PDMS was mixed with the MWCNT/Ethanol solution. The ethanol was completely removed after 5 h at 80 °C and the printable ink is ready. Then, using a 3D printing process, the capillary pressure sensor was fabricated. At 1 kHz of frequency and under pressure, the capacitance variation was measured, respectively, at the initial state, 30 Pa, 90 Pa, and 120 Pa to get values of 0.08 nF, 1.50 nF, 3.56 nF, and 4.92 nF respectively. A high performance achieved for example for sensitivity higher than 547.9 $kPa^{-1}$ and high durability after 500 pressing cycles. Pyo et al. (2017) proposed a new structure for a sensitive pressure sensor using CNT material, where the CNTs are designed as a coated polyester fabric. The multi-walled carbon nanotubes with 20 nm of a diameter and 10 μm of a length, were dispersed in deionized water DI by applying Carbon Nano Technology. Then, a commercially polyester fabrics are dipped into the dispersed solution and rapidly coated with MWCNT due to its absorption property to be finally dried in an oven at 120 °C. After the fabrication of the pressure sensors, many studies were done to prove that the range of the variable resistance can be controlled by modulating the dipping number. Under pressure, the nanocomposite sensor shows high electrical properties. An external force is applied onto the coated polyester fabric the contact between CNT decrease resulting in increasing the conductive pathways, consequently, the decrease of resistance. When pressure is applied, the sensor shows two linear measurements. Between 0 to 10 kPa, the resistance varies due to the contact of CNT fibers. Once the pressure exceeds 10 kPa, the increase of fibers contact is detected owing to the increase of resistance. Hence, a maximum sensitivity was measured by 10.63%/kPa for a pressure of 10 kPa.

Maddipatla et al. (2017) developed a capacitive sensor-based CNT and Polydimethylsiloxane (PDMS), which is designed using screen printing. A dielectric layer was integrated between plates based on PDMS polymer. Then, the CNT electrodes are printed using CNT ink on PDMS layer with a thickness of 8.9 μm and

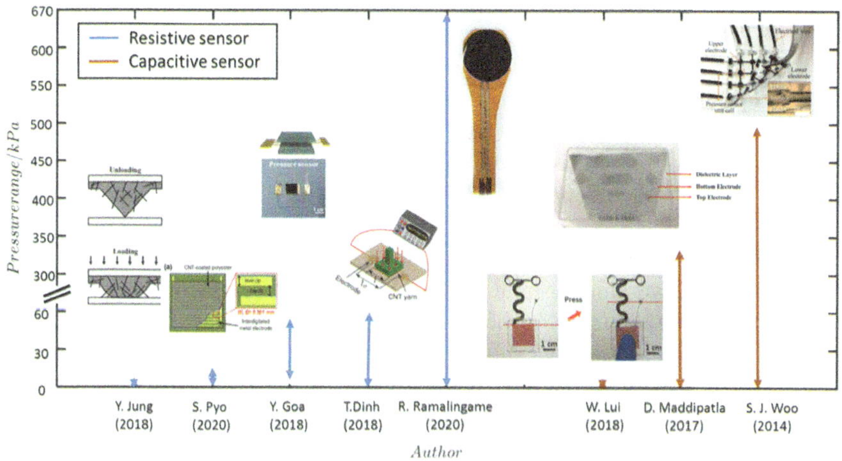


**Fig. 2** Capacitive and resistive pressure sensors overview

average roughness of 0.8 μm. The electrode was directly screen printed using conductive CNT ink onto the PDMS. From experimental results, the capacitance range variation was defined from 6.49 pF to 7.02 pF for a maximum pressure subjected by 337 kPa. The percentage of change of the capacitance compared with the based capacitance value was estimated at 8.2%. A 0.021% change in capacitance per kPa and a correlation coefficient of 0.9971 was also determined for the CNT-based pressure sensor. Woo et al. (2014) present a highly elastic capacitive pressure sensor using a combination of soft-lithographic replication and contact printing. The printing ink is a mixture of 10% MWCNT dispersed on PDMS liquid, which was diluted with the toluene. The sensor was composed of 16 individual capacitively-coupled pressure sensing cells for increasing the pressure surface detection. Each cell is designed with two MWCNT/PDMS sensing electrodes separated by ecoflex layer as a dielectric layer. When no pressure applied, the initial capacitance was defined to be 2.66 pF, under pressure increases to attend 3.18 pF with a relatively small stander deviation of 0.36 pF for 0.5 MPa of subjected pressure (Fig. 2).

Ramalingame et al. (2019) developed a high-pressure sensor based on the multi-walled carbon nanotube material, where the sensitive force/pressure information is required. Therefore, a polydimethylsiloxane (PDMS) composite is used as a polymer for MWCNT dispersion. With a non-complex fabrication process, 1 wt%.MWCNT was mixed with PDMS polymer using tetrahydrofuran (THF) solvent as the dispersion medium. The pressure sensor exhibits a high sensing performance due to its fast response to detected low pressure, high durability after 100 cyclic loading/unloading test and high sensitivity up to 670 kPa. A measurement system for hand movement detection was proposed with high sensing capabilities of the fabricated pressure sensor.

## 2.2 *Strain Sensors*

### 2.2.1 Sensor Principle

A specific type of piezoresistive device that can be fabricated based on CPCs and in particular on carbon nano-filled polymer composites is the mechanical strain sensor. Its sensing principle relies on the modification of the percolated network upon mechanically stressing or stretching the composite. Figure 3 shows the principle of a nanocomposite working as strain sensor. Mechanical deformation is applied to the composite causing an impact on the structure of the percolated network (here shown as a 1-dimensional linkage for explanatory purposes) which in turn changes the electrical resistance of the composite. As illustrated in Fig. 3, the percolated network is extended by the mechanical deformation and the contact distance between neighbouring conductive particles is increased or interrupted (red circles in Fig. 3). By that, the distance between neighboured CNTs increases and may surpass the tunnelling distance (Hu et al. 2010).

This phenomenon was already reported for polymer/CNT sensors subjected to unidirectional elongation (Hu et al. 2008). According to Hu et al. the increase of distances between rod-like particles is an important factor affecting the electrical resistance change in polymer/CNT strain sensors (Pham et al. 2008). At low deformation (within the linear viscoelastic range) this process is reversible. In most cases, the plastic regime of a thermoplastic composite should not have been reached upon mechanical load. This is desirable for applications in which the developed sensor material is intended to be used repeatedly. In general, while doing a piezoresistive test the changes in electrical resistance are measured with time and afterwards correlated to the applied mechanical deformation. Once the characteristic electrical response of the composite is known upon mechanical deformation, it can be correlated to a defined strain. In order to transduce from the electrical response to mechanical deformation a transforming function is needed. A linear case, where the change in electrical resistance is directly proportional to the mechanical strain, can be denoted by the following equation:

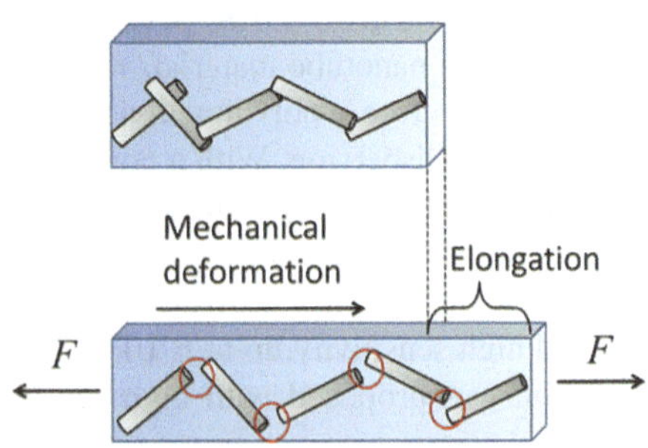


**Fig. 3** Electrical conductive CNT network in a polymer matrix under applied strain

$$S_{GF}\,\varepsilon = \frac{\Delta R}{R_0} \tag{4}$$

$$\Delta R = R_{measured} - R_0 \tag{5}$$

where $S_{GF}$ is the strain gage factor, $\varepsilon$ is the mechanical strain, $R_0$ is the electric resistance previous to load application.

The $S_{GF}$ is the coefficient that correlates the strain and the resistance changes and it can be used to estimate the strain from a measured resistance. However, Eq. 4 is only applicable for cases or deformation ranges where the deformation upon applied load is linear. Elastomers for example have a non-linear behaviour. For these cases the resulting curve can be fitted to a polynomial function from which the strain can be obtained (Bautista-Quijano et al. 2013). Moreover, different gage factors can be obtained depending on the deformation behavior of a composite upon loading. For instance, Ku et al. found that two gage factors can be adjusted to fit a nonlinear piezoresistive behaviour, where each follows the strain in the elastic and plastic zone, respectively (Ku-Herrera et al. 2013).

### 2.2.2 Recently Developed Strain Sensors: An Overview

Wajahat et al. (2018) developed a flexible strain sensor for Human-machine interface and health monitoring requirements. Using a meniscus-guided printing system, the sensor was fabricated based on multi-walled carbon nanotubes (MWCNT) and polyvinylpyrrolidone (PVP) polymer. To ensure high-performance measurement, a cycle test during 1500 bending cycles was done at 2.0% of strain to demonstrate the high stability response. In the same context, a gauge factor was estimated by 12.87 under compressive strain and 13.07 under tensile strain due to the MWCNT/PVP reproducible response. When strain applied, a rearrangement contact of the MWCNTs was induced resulting in a change of the tunnelling distance between them, consequently, a variation of the ($\Delta$R/R0) resistance of the printed sensor. To exhibit high performance of the sensor, finger glove was fabricated for detection of the fingers movements and remote control of robotic equipment. Ko et al. (2018) designed a conductive textile via vacuum-filtrated where the MWCNT powder was used as a conductive material dispersed in deionized water (DI) solution. The investigation of the fabricated conductive textile was done based on its resistance variation. Many studies have been conducted to ensure the stability of the sensor. Precisely, a stretchability cyclic test was done at a strain rate of 0–20%. The nanocomposite material shows a uniform resistance variation less than ±3% during 10.000 cycles of stretching. To enhance the performance of the MWCNT/DI sensor, a motion-sensing glove was proposed for the interaction with the human skin by the fingers monitoring. Amjadi et al. (2014) suggest a sandwich structure of a new strain sensor. Herein, highly stretchable polymer elastomer wrapped the nanocomposite of silver nanowire network (AgNW) using PDMS elastomer as a dispersed solution in the form of a sandwich structure (the AgNW thin film embedded between two layers of PDMS polymer). Due to the high AgNW/PDMS interaction and the silver nanowires

**Table 1** Developed nanocomposite strain sensors and reported applications

| Reference | Nano-particles | Polymer | Fabrication technology | Application |
|---|---|---|---|---|
| Wajahat et al. (2018) | MWCNT | PVP | Meniscus-guided printing | Robotic hand control |
| Ko et al. (2018) | MWCNT | DI | Vacuum-filtrated | Conductive textile glove |
| Amjadi et al. (2014) | CNT | Silicon | Wet-spinning assembly | Woven glove/Wristlet |
| Tang et al. (2018) | AgNW | PDMS | Sandwich structure | Smart glove/Avatar robot |

density, a strong piezoresistivity with tunable gauge factors was estimated between 2 and 14 with a 70% of a stretchability, which is noted for high stretchable and wearable sensing response. A higher Young's modulus of AgNWs was calculated between 81 to 176 GPa. The sensor response was studied by the 3D resistor network behavior where the variation of resistance under strain was detected. Based on the experimental and simulation results, a smart glove was developed by assembling five nanocomposite sensors for detection of the fingers movements and the control of an avatar robot. Tang et al. (2018) proposed a coaxial wet-spinning assembly approach for developing a highly sensitive nanocomposite sensor. Using a pure silicon elastomer as a polymer, the multi-walled carbon nanotubes are dispersed into the prepared polymer solution. A core-sheath fibre strain sensor is fabricated based on the MWCNT/Silicon interaction. The carbon nanotube polymeric composite core of the stretchable fiber is surrounded by an insulating sheath, similar to conventional cables and shows excellent electrical conductivity with a low percolation threshold (0.74 vol %). The core-sheath fiber strain sensor shows a great performance by the means of the high stretchability percentage about 300%, high reproducibility with more than 10.000 cycles, great sensitivity by a large gauge factor range estimated on 1378 and Young's modulus approximated equal to 0.49 MPa. To explore the application for the human motion detection, a commercial glove is developed where the CSF sensor was attached to investigate the real response corresponding to the wrist and human fingers movements (Table 1).

## 3 Hybrid Glove with Implemented Pressure and Strain Sensors

A real-time feedback system that can offer rehabilitation system for the patients with fingers and wrist disabilities is presented. The proposed setup is a flexible, portable, cost-effective home rehabilitation system, and it consists of a regular glove

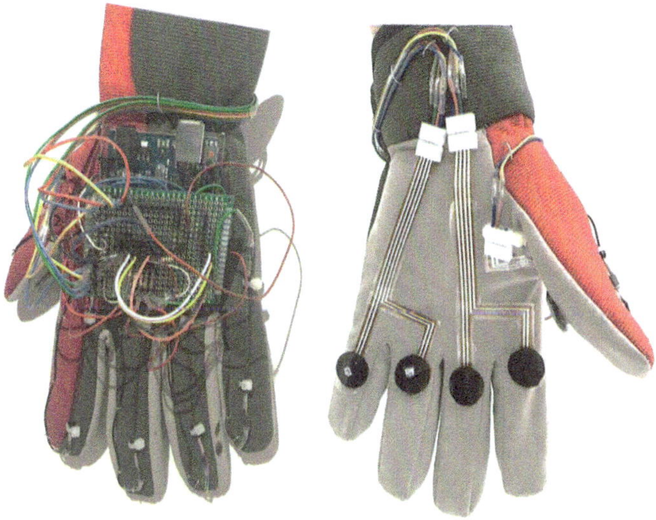

**Fig. 4** Developed glove with attached strain and pressure sensors

mounted with a nanocomposite filament-based strain sensor to monitor the individual finger movements and pressure sensors, that can monitor the pressure distribution in the palm (Atitallah et al. 2019). The nanocomposite fiber strain sensor is placed in each finger, and pressure sensors are placed inside the hand. Sensor placement and the number of sensors depend on the finger movement and maximum pressure distribution formed during the handling of the object and this can be altered depends on patients' need as well as doctor's advice (Fig. 4).

### *3.1 SBS/C-TPU Strain Sensor*

SBS (styrene-butadiene-styrene rubber) and the C-TPU (conductive thermoplastic urethane) were weighted according to the desired composition and mixed in a glass flask (Ramalingame et al. 2017c). This mixture is then added to an extruder device were the pellets are melted at a temperature up to 250 °C under a constant pressure exerted by the screw rotation. The mixture is pushed towards a heated metal die where the filaments form by passing through the die hole (Torres et al. 2018). For this study, a 1 mm die hole is used, resulting in filaments of 1 mm in diameter. The resulting composite is a filament blend made of SBS and C-TPU. 12 cm parts were separated, and metal clamps were attached as observed in Fig. 1. The initial resistance of obtained filaments was approximately 1 MΩ and obtained a gauge factor (GF) equal to 24.

Several tests were carried out to determine the characteristics of the behaviours related to the stretching operation including mechanical test, Hysteresis test, and

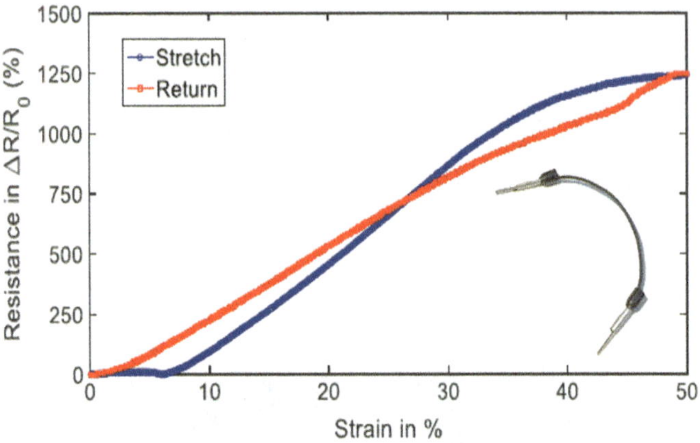


**Fig. 5** Hysteresis behaviour under applied strain-release cycle

cyclic test. Figure 5 shows that the electrical resistance change increases upon increasing strain and decreases upon load release. For instance, for a strain of 50% the increase in electrical resistance change is as high as 1200%, which is indicative of large sensitivity. The apparent low magnitude of resistance change corresponding to low strain ($\geq$5%) implies that the percolated network inside the fiber is not greatly modified by small amounts of stretching, yet the change in resistance is as high as $\sim$10%. In order to study the behaviour of the fibers upon cyclic straining, loading-unloading cycles were performed up to 60 mm of elongation as shown in Fig. 6. The electrical response of the filament is able to follow the applied cyclic strain. The filament shows an increase in maximum electrical resistance reached with each cycle and results in higher values than the previous loading cycle until the 18th cycle where the resistance value reached in further cycles remains the same. This can be due to a re-accommodation of the polymer chains after releasing the load which in turn modifies the way the MWCNTs are initially arranged inside the polymer. Nevertheless, in all sensors evaluated the maximum and minimum electrical resistances remain similar after a specific amount of cycles. This shows that higher repeatability in the electrical response of the sensors is achieved once a determined amount of cycles is reached. This behaviour can be due to CNTs re-accommodation occurring around deformed elastomeric thermoplastic polymer chains at each stretching-shrinking step until no more CNT re-accommodation takes place along the deformed polymer chains. This in practical terms can be seen as a 'training' step in which the sensor has to be subjected to a pre-defined amount of pre-stretching cycles to secure a repeatable response and thus reducing the hysteresis.

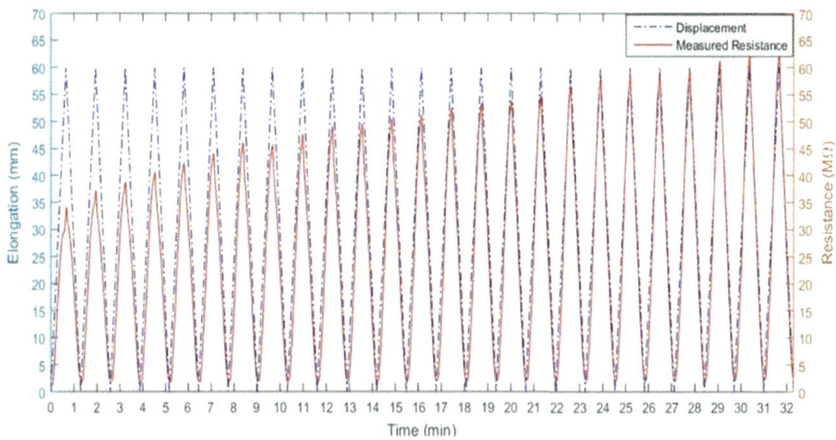


**Fig. 6** Displacement and resistance measured on SBS/CTPU filaments upon cyclic evaluation

## 3.2 CNT/PDMS Pressure Sensor

A high-pressure sensor was developed based on the multi-walled carbon nanotube material (MWCNT), where the sensitive force/pressure information is required. Therefore, a polydimethylsiloxane (PDMS) composite is used as a polymer for MWCNT dispersion. The fabrication process of the nanocomposite pressure sensor is carried out in accordance with our prior work (Ramalingame et al. 2017a). MWCNTs in the concentration of 1 wt.% are pre-dispersed in an organic solvent (Ramalingame et al. 2017b) using sonication and magnetic stirring process and later dispersed in PDMS with higher amplitudes of sonication and magnetic stirring. Later this nanocomposite dispersion is mixed with a cross-linking agent to facilitate the curing process of the polymer which is complete in 2 h under a temperature of 120 °C. The fabricated nanocomposite is then attached to a silver-based circular interdigital electrode for electrical readout and can be utilized as either piezoresistive (Ramalingame et al. 2019) or piezo-capacitive (Ramalingame et al. 2017a) pressure sensor based on the excitation signal and desired measurement signal (see Fig. 7). This module exhibits a high sensing performance due to its fast response to detect low pressure, high durability and high sensitivity up to 670 kPa. The sensor response was investigated using a cyclic test after 100 cyclic loading/unloading. The resistance converges approximately to the same range of values for every load cycle. Figure 8 illustrates the repetitive behavior of the sensor variation, where the resistance varies from 1 kΩ for low-pressure values to more than 15 kΩ in high applied pressure values.

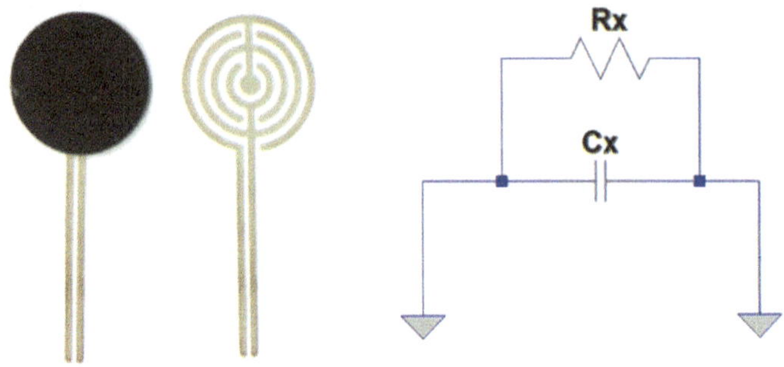


**Fig. 7** **a** Nanocomposite sensor based on MWCNT/PDMS polymer, **b** Equivalent circuit of the pressure sensor

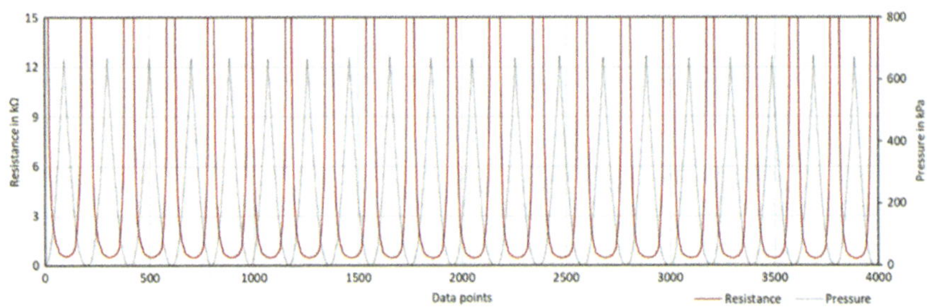


**Fig. 8** Resistance variation under applied cyclic pressure test

## 3.3 Data Acquisition System

Based on the measured resistance value of each pressure/strain sensor unit, the corresponding pressure/strain value can be calculated out. Practically, sensors are divided into two groups. Each sensor group is controlled by one corresponding multiplexer. Under the controlling information sent from the Arduino board, each multiplexer will active only one channel at a time, and this activated channel is connected to one sensor unit. This sensor unit together with the reference resistor $R_{ref}$ then forms a voltage divider charged by 5 V. An operational amplifier as voltage follower is used to isolate the Arduino ADC out of the measurement circuit. The activated sensor resistance value can be calculated as:

$$R_{Sensor} = \frac{R_{ref}\, V_{out}}{5 - V_{out}} \tag{6}$$

where $R_{ref}$ is the given reference resistance value and $V_{out}$ can be acquired by the Arduino ADC. Hence, the resistance value of every sensor can be calculated one by one. the piezo-resistive behavior of the nanocomposite sensor was considered. Therefore, due to the increase in the number of conducting paths in the resistance

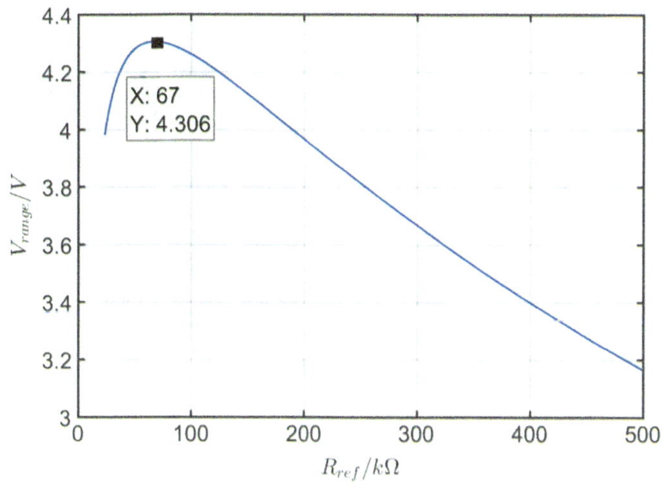


**Fig. 9** Sensor voltage variation range based on the reference resistance vector

network of the nanocomposite, the resistance will be decreased. The effect of tunneling resistance decreases while the current-carrying increases. For accurate and high-resolution measurements, the choice of the appropriate resistance value is critical. It enables to cover the maximum range of the output sensor voltage $V_s$, taking into account the range of the sensor resistance variation $R_{sensor}$. For this, a Matlab algorithm is developed to calculate the maximum range of voltage variation of the sensor. The defined resistance range was assumed between 5–500 kΩ where the reference resistance should be calculated to obtain the maximum voltage sensor range. An iterative computing algorithm is implemented, where for each value of the $R_{ref}$ vector, the sensor voltage Vs is calculated by applying Eq. (7)

$$Vs = Vcc \frac{R_{sensor}}{R_{sensor} + R_{ref}} \tag{7}$$

based on all other remaining sensor resistance vector to extract the maximum range of $V_s$ for each round.

At the end of this algorithm, the optimal value of $R_{ref}$ is obtained by considering the maximum range value from the maximum range vector. As illustrated in Fig. 9, the maximum output voltage range of $V_s$ is expected for the measurement range between 5–500 kΩ, where the maximum Vs range equal to 4.3063 V, corresponds to 67 K as a value of the reference resistance $R_{ref}$. The calculated resistance value of each sensor unit will be sent by the Arduino board through serial communication protocol to the GUI software written in Python on PC.

For the pressure sensor units, their resistance value decreases when the given pressure increases (from Fig. 10a, color yellow presents light pressure while blue color

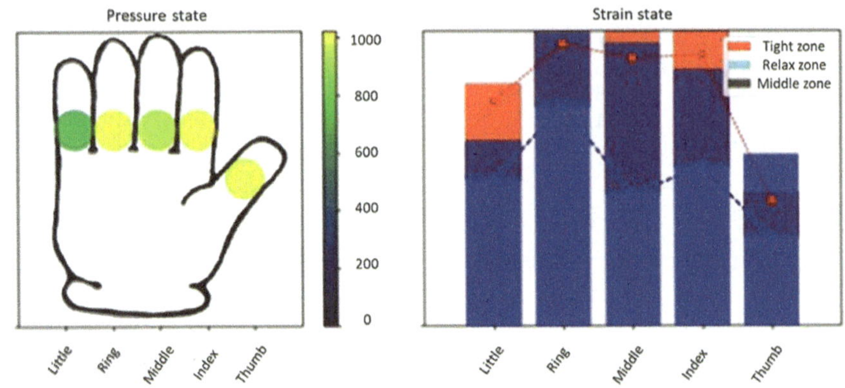


**Fig. 10** Measurement result in the software system

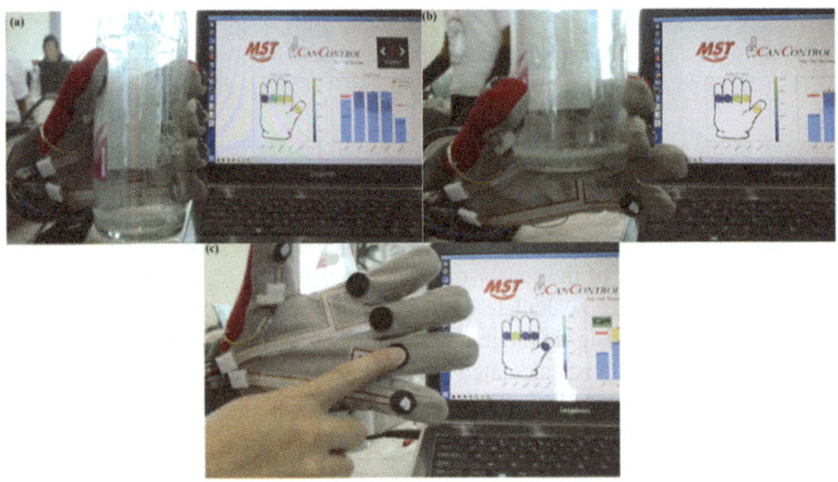


**Fig. 11** Practical experiments: **a** Holding the bottle with all fingers, **b** Holding the bottle with thumb, fore and middle finger **c** Pressing the individual sensor

means great pressure). For the strain sensor units, their resistance value increases when the strain increases (as the histogram in Fig. 10b).

In order to evaluate the working of the glove, we have recorded the movement of the hand while holding a glass bottle. From Fig. 11, it can be inferred that while holding a bottle, the pressure range is in the order: thumb, fore finger, middle finger, ring finger, babby finger. Figure 11b shows the pressure distribution while holding the object with fore and middle finger. Individual sensors were evaluated which can be seen in Fig. 11c.

## 4 Conclusion

Among several intelligent nanoscale materials, carbon nanotubes (CNTs) show an interesting behavior in the nanocomposite matrix based on their remarkable electrical, mechanical, and electrochemical properties. In this context, a deep study was carried out on polymer carbon nanotube composite based pressure/strain sensors as well as their properties and working principles. Besides, nanocomposite sensors present a highly accurate measurement unit in biomedical and telemanipulation applications due to their flexibility and human body compatibility. In this paper, we reported the design and implementation of a safe, flexible and easily transportable electronic glove that can offer multiple hand movement possibilities for the recovery process of the patients with hand-related problems. It consists of a hybrid system including strain and pressure sensors for force distribution monitoring and gesture tracking respectively.

**Acknowledgements** This work is funded by the Deutsche Forschungsgemeinschaft (DFG, German Research Foundation) - Project-ID 416228727 - SFB 1410, applicant Prof. Dr.-Ing. Olfa Kanoun. Besides, the authors would like to thank the German Academic Exchange Service "DAAD" (Deutscher Akademischer Austauschdienst) for their support through the exchange project Bismon-57477606.

## References

Amjadi, M., Pichitpajongkit, A., Lee, S., Ryu, S., & Park, I. (2014). Highly stretchable and sensitive strain sensor based on silver nanowire-elastomer nanocomposite. *ACS Nano*, *8*(5), 5154–5163.

Atitallah, B. B., Ramalingame, R., Derbel, N., & Kanoun, O. (2019). Muscle movement tracking using nanocomposite based pressure sensor. In *2019 16th International Multi-conference on Systems, Signals & Devices (SSD)* (pp. 483–488). IEEE.

Bautista-Quijano, J., Aviles, F., & Cauich-Rodriguez, J. (2013). Sensing of large strain using multiwall carbon nanotube/segmented polyurethane composites. *Journal of Applied Polymer Science*, *130*(1), 375–382.

Boian, R., Sharma, A., Han, C., Merians, A., Burdea, G., Adamovich, S., et al. (2002). Virtual reality-based post-stroke hand rehabilitation. *Studies in health technology and informatics* (pp. 64–70).

Burtney, M. (2020). Keep facts handy: 5 simple statistics to motivate hand protection. https://www.ishn.com/articles/107351-keep-facts-handy-5-simple-statistics-to-motivate-hand-protection.

Dinh, T., Nguyen, T.-K., Phan, H.-P., Fastier-Wooller, J., Tran, C.-D., Nguyen, N.-T., et al. (2018). Electrical resistance of carbon nanotube yarns under compressive transverse pressure. *IEEE Electron Device Letters*, *39*(4), 584–587.

Dinh, T. N., Sowade, E., Arreba, A., Belau, R., Blaudeck, T., Baumann, R. R., et al. (2011). Performance of liquid-deposited multiwalled carbon nanotube films under strain. *Printing Future Days*, 109–115.

Dovat, L., Lambercy, O., Gassert, R., Maeder, T., Milner, T., Leong, T. C., et al. (2008). Handcare: A cable-actuated rehabilitation system to train hand function after stroke. *IEEE Transactions on Neural Systems and Rehabilitation Engineering*, *16*(6), 582–591.

Ergen, H. I., & Oksuz, C. (2019). Evaluation of load distributions and contact areas in 4 common grip types used in daily living activities. *The Journal of Hand Surgery*.

Gao, Y., Yu, G., Tan, J., & Xuan, F. (2018). Sandpaper-molded wearable pressure sensor for electronic skins. *Sensors and Actuators A: Physical, 280*, 205–209.

Hu, N., Karube, Y., Arai, M., Watanabe, T., Yan, C., Li, Y., et al. (2010). Investigation on sensitivity of a polymer/carbon nanotube composite strain sensor. *Carbon, 48*(3), 680–687.

Hu, N., Karube, Y., Yan, C., Masuda, Z., & Fukunaga, H. (2008). Tunneling effect in a polymer/carbon nanotube nanocomposite strain sensor. *Acta Materialia, 56*(13), 2929–2936.

Jung, Y., Shin, M. G., Tak, H. J., Jung, K. K., & Ko, J. S. (2018). Flexible pressure sensor made using PDMS containing carbon nanotubes. In *2018 IEEE Micro Electro Mechanical Systems (MEMS)* (pp. 501–503). IEEE.

Kanoun, O., Müller, C., Benchirouf, A., Sanli, A., Bouhamed, A., Al-Hamry, A., et al. (2014). Potential of flexible carbon nanotube films for high performance strain and pressure sensors. Nanotechnology for Optics and Sensors (One Central (OCP)).

Kanoun, O., Bouhamed, A., Ramalingame, R., Bautista-Quijano, J. R., Rajendran, D., & Al-Hamry, A. (2021). Review on conductive polymer/CNTs nanocomposites based flexible and stretchable strain and pressure sensors. *Sensors, 21*(2), 341; 1–29.

Karime, A., Al-Osman, H., Gueaieb, W., & El Saddik, A. (2011). E-glove: An electronic glove with vibro-tactile feedback for wrist rehabilitation of post-stroke patients. In *2011 IEEE International Conference on Multimedia and Expo* (pp. 1–6). IEEE.

Ko, J., Jee, S., Lee, J. H., & Kim, S. H. (2018). High durability conductive textile using MWCNT for motion sensing. *Sensors and Actuators A: Physical, 274*, 50–56.

Ku-Herrera, J., Avilés, F., & Seidel, G. (2013). Self-sensing of elastic strain, matrix yielding and plasticity in multiwall carbon nanotube/vinyl ester composites. *Smart Materials and Structures, 22*(8).

Lenhoff, H. M., Wang, P. P., Greenberg, F., & Bellugi, U. (1997). Williams syndrome and the brain. *Scientific American, 277*(6), 68–73.

Liu, W., & Yan, C. (2018). Direct printing of stretchable elastomers for highly sensitive capillary pressure sensors. *Sensors, 18*(4), 1001.

Maddipatla, D., Narakathu, B. B., Ali, M. M., Chlaihawi, A. A., & Atashbar, M. Z. (2017). Development of a novel carbon nanotube based printed and flexible pressure sensor. In *2017 IEEE Sensors Applications Symposium (SAS)* (pp. 1–4). IEEE.

Mulas, M., Folgheraiter, M.,& Gini, G. (2005). An emg-controlled exoskeleton for hand rehabilitation. In *9th International Conference on Rehabilitation Robotics, 2005. ICORR 2005* (pp. 371–374). IEEE.

Park, S., Meeker, C., Weber, L. M., Bishop, L., Stein, J., & Ciocarlie, M. (2018). Multimodal sensing and interaction for a robotic hand orthosis. *IEEE Robotics and Automation Letters, 4*(2), 315–322.

Pham, G. T., Park, Y.-B., Liang, Z., Zhang, C., & Wang, B. (2008). Processing and modeling of conductive thermoplastic/carbon nanotube films for strain sensing. *Composites Part B: Engineering, 39*(1), 209–216.

Pyo, S., Jo, E., Kwon, D.-S., Kim, W., Chang, W., & Kim, J. (2017). Fabrication of carbon nanotube-coated fabric for highly sensitive pressure sensor. In *2017 19th International Conference on Solid-State Sensors, Actuators and Microsystems (TRANSDUCERS)* (pp. 962–965). IEEE.

Ramalingame, R., Chandraker, P., & Kanoun, O. (2017a). Investigation on the influence of solvents on MWCNT-PDMS nanocomposite pressure sensitive films. In *Multidisciplinary Digital Publishing Institute Proceedings* (Vol. 1, p. 384).

Ramalingame, R., Hu, Z., Gerlach, C., & Kanoun, O. (2017b). Shoe insole with MWCNT-PDMS-composite sensors for pressure monitoring. In *2017 IEEE SENSORS* (pp. 1–3). IEEE.

Ramalingame, R., Hu, Z., Gerlach, C., Rajendran, D., Zubkova, T., Baumann, R., et al. (2019). Flexible piezoresistive sensor matrix based on a carbon nanotube PDMS composite for dynamic pressure distribution measurement. *Journal of Sensors and Sensor Systems, 8*(1), 1.

Ramalingame, R., Torres, R., Hu, Z., Chandraker, P., & Kanoun, O. (2017c). Electrical impedance analysis on nanocomposite based pressure and strain sensor. In *10th International Workshop on Impedance Spectroscopy (IWIS)* (pp. 102–103).

Stilli, A., Cremoni, A., Bianchi, M., Ridolfi, A., Gerii, F., Vannetti, F., et al. (2018). Airexglove-a novel pneumatic exoskeleton glove for adaptive hand rehabilitation in post-stroke patients. In *2018 IEEE International Conference on Soft Robotics (RoboSoft)* (pp. 579–584). IEEE.
Tang, Z., Jia, S., Wang, F., Bian, C., Chen, Y., Wang, Y., et al. (2018). Highly stretchable core-sheath fibers via wet-spinning for wearable strain sensors. *ACS Applied Materials & Interfaces, 10*(7), 6624–6635.
Torres, R., Cheng, Z., Ramalingame, R., & Kanoun, O. (2018). Electrical characterization of elongation sensors based on SBS-CTPU filaments. In *2018 15th International Multi-Conference on Systems, Signals & Devices (SSD)* (pp. 1212–1215). IEEE.
Wajahat, M., Lee, S., Kim, J. H., Chang, W. S., Pyo, J., Cho, S. H., et al. (2018). Flexible strain sensors fabricated by meniscus-guided printing of carbon nanotube-polymer composites. *ACS Applied Materials & Interfaces, 10*(23), 19999–20005.
Wang, Z., & Ye, X. (2013). A numerical investigation on piezoresistive behaviour of carbon nanotube/polymer composites: mechanism and optimizing principle. *Nanotechnology, 24*(26).
Williams, R., Farquharson, L., Palmer, L., Bassett, P., Clarke, J., Clark, D. M., et al. (2016). Patient preference in psychological treatment and associations with self-reported outcome: national cross-sectional survey in England and wales. *BMC Psychiatry, 16*(1), 4.
Woo, S.-J., Kong, J.-H., Kim, D.-G., & Kim, J.-M. (2014). A thin all-elastomeric capacitive pressure sensor array based on micro-contact printed elastic conductors. *Journal of Materials Chemistry C, 2*(22), 4415–4422.

# Wireless Body Sensor Networks with Enhanced Reliability by Data Aggregation Based on Machine Learning Algorithms

**Mbarka Belhaj Mohamed, Amel Meddeb-Makhlouf, Ahmed Fakhfakh, and Olfa Kanoun**

**Abstract** Wireless Body Sensor Networks (WBSNs) are widely used in Internet of Things (IoT) based health care technologies, where the health status of patients can be monitored through a group of small-powered and lightweight sensor nodes. Energy consumption is thereby a major issue. One of the reasons for energy losses in WBSN are retransmission process, when data collision takes place or data are not received properly due to channel fading. In order to reduce the necessity for data retransmission, we propose to reduce the transmitted data by applying data aggregation techniques. This raises also the network lifetime by minimizing the resources consumption of the sensor nodes. Nevertheless, it may degrade significantly the service quality metrics, such as data reliability and communication security. In this chapter, an accurate data classification model for multiple health signals based on different machine learning algorithms is proposed to ensure the reliability of data aggregation. The performance evaluation shows that the Random Forest algorithm is the best classifier in terms of accuracy (97%) and sensitivity (92%) under general conditions.

**Keywords** Accuracy enhancement · Classification · Machine learning · Reliability · Data aggregation · WBSN

---

M. Belhaj Mohamed (✉) · A. Fakhfakh
Laboratory Signals, Systems, Artificial Intelligence and Networks (SM@RTS), National School of Engineers of Gabes, Gabes, Tunisia
e-mail: mbarkaenig@gmail.com

A. Fakhfakh
e-mail: ahmed.fakhfakh@enetcom.usf.tn

A. Meddeb-Makhlouf
Research Unit New Technologies and Telecommunications Systems, National School of Electronics and Telecommunications of Sfax, Sfax, Tunisia
e-mail: amel.makhlouf@enetcom.usf.tn

O. Kanoun
Measurement and Sensor Technology, Chemnitz University of Technology, Chemnitz, Germany
e-mail: olfa.kanoun@etit.tu-chemnitz.de




O. Kanoun and N. Derbel (eds.), *Advanced Sensors for Biomedical Applications*, Smart Sensors, Measurement and Instrumentation 38,
https://doi.org/10.1007/978-3-030-71225-9_4

# 1 Introduction

Nowadays, Internet of Things (IoT) is supporting advancing digitalization in several sectors. Many objects in daily life become thereby part of the internet supported by communication and computing capabilities. In healthcare technologies, IoT involves numerous types of inexpensive sensors, wearables and implants. In fact, persons enjoy current medical healthcare services at everyplace, any time, and it improved thereby their quality of life. A Wireless Body Sensor Network (WBSN) consists of several sensors (Bahae et al. 2020) and represents a special case of a Wireless Sensor Network (WSN). Figure 1 shows the WBSN based system for personal health care.

Each stage of this architecture is further explained in more detail as follows:

(1) The Intra-BSN (Tiers 1 in Fig. 1): In this stage, some sensors are placed on the human body as (wearable sensors) or implanted under the skin as (in-body sensors). Different kind of sensor can be used and integrated such as electroencephalogram (EEG), electrocardiogram (ECG), electromyography (EMG), or blood pressure (BP) measuring sensors. Therefore, the recorded data are sent to the local processing unit (LPU) like: a smart phone, a tablet, or a computer through wireless medium, and then the LPU transfers the data to the next stage.
(2) The Inter-BSN (Tiers 2 in Fig. 1): In this stage, the data transfer from Tiers 1 to Tiers 2 over one of the selected wireless communication technologies which could be access point, Wi-Fi, or cellular base station. Collected data from this stage need to be forwarded to Tiers 3.
(3) The Beyond BSN (Tiers 3 in Fig. 1): In this stage, the data is stored and analyzed to make final decision about the health status. For that, data may be transferred to doctors or hospitals and to intermediate family. In fact, received data should be evenly accurate and reliable as required.

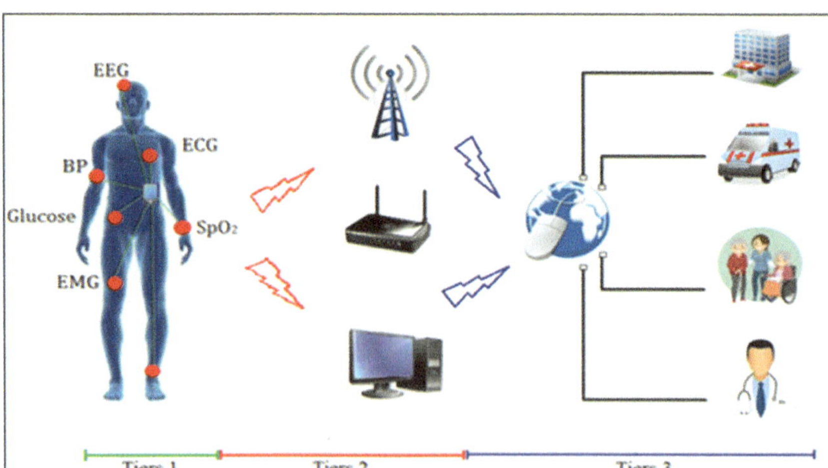


**Fig. 1** WBSN based system for personal health care

From recent study, the energy usage of each biosensor node and the data transmission reliability present a huge significance on WBSNs (Mehrani et al. 2020; Sabeeh Hmoud Altamimi et al. 2020). The problem becomes more complicated in wireless sensor networks with many nodes, where neighboring sensor nodes often have overlapping sensing ranges. That being the case, they sense the same phenomenon which results in production of large volumes of redundant data. To reduce the amount of data transmission, data redundancy is eliminated at intermediate sensor nodes by performing data aggregation (Abbasian Dehkordi et al. 2020). Indeed, data aggregation protocols that consider compromised nodes need improvements of reliability (Kunal Kishan et al. 2020).

In this context, a WBSN needs a high degree of reliability as this immediately affects the quality of patient monitoring (Tarique et al. 2020). A prime requirement is that the health care experts receive the monitored data correctly. For this reason, we enhance the data reliability at the receiving side that is meaning at Tiers 3. This chapter surveys the well known supervised machine learning algorithms used for data classification of ECG, EMG and BP aggregated data. The main objective thereby is the development of advanced and efficient decision making model for cardiovascular, muscle contraction and hypertension diseases. Due to the sensitivity and criticality of the data carried and handled by WBSN, the accuracy is a critical issue and widely discussed.

This chapter is structured as follows: In Sect. 2 a study of sate of the art is provided. Then, the proposed methods are explained in Sect. 3, the simulations and results are discussed in Sect. 4. Discussions relating to the implementation of the applied model is given in Sect. 5. Finally, conclusion and remarks are drawn in the last section.

## 2 Related Works

Detecting abnormal medical data is pertinent to the life of the monitored patient (Belhaj Mohamed et al. 2018, 2019, 2020). Over these years, a huge number of healthcare applications based on machine learning algorithms for improving decision making have been proposed.

As stated in the reports published by several organizations, it may be seen that about 50 million people are at risk of cardiovascular diseases around the world (Verma et al. 2020). Furthermore, various types of heart diseases are the major usual causes of mortality. Hence, the proposed work investigates an electrocardiogram (ECG) classifying strategy using many techniques. The pre-processing process of the proposed method is based on three essential steps as flows: removing ECG interference, peak points detection and feature extraction. In fact, their goal is to classify the ECG signal database into normal or abnormal signal. For the classification process, GWO-MSVM, SVM, Adaboost, ANN and Naive Baye are used in order to extracted feature classification. The obtained precision of these algorithms is 99.9%, 94%, 93%, 87.57% and 85.28% respectively.

Even though a huge research study for heart rate disease. In Plawiak et al. (2020) new and efficient methods are introduced to detect myocardium dysfunctions. They used 744 fragments of ECG signal database related to 29 patients from the MIH-BIH Arrhythmia database. The suggested research using a new evolutionary-neural system, based on the SVM classifier. The performance evaluation of the applied method shows a high sensitivity (90.19%), specificity (99.39%), and accuracy (98.85%) compared to the existing methods.

Many observations judge that the blood pressure measurements and prediction is a significant factor for cardiovascular diseases. For that, it should be preserved under regular control. The used data set by Alghamdi et al. (2020) is the oscillometric wave forms of individuals using a cuff. These wave forms are divided on three periods which are: the first period from the starting point to the systolic blood pressure (SBP), the second period is between systolic blood pressure (SBP) and diastolic blood pressure (DBP), and the third period is between diastolic blood pressure (DBP) and the end of the waveform. For the estimation of SBP values using the kNN algorithm, weighted kNN, and Bagged Trees, the obtained Mean Absolute Error (MAEs) are 3.590, 3.520, and 4.499, respectively and concerning the estimation of DBP, the calculated MAEs are 11.077, 11.032, and 13.069, respectively.

It is necessary to focus on the major causes of the hypertension status. Several research prove that it is associated with various risk factors and can be treated by lifestyle modifications and medications. The objective (Alkaabi et al. 2020) nowadays is to yield a hypertension predictive model without the need of invasive clinical procedures. As a reason that the machine learning provides the chance of having a rapid predictive, authors evaluated several algorithms based on different patients attributes for instance: age, gender, education level, employment, tobacco use, physical activity, abdominal obesity, history of diabetes, history of high cholesterol, etc.

All algorithms showed more or less similar performances: random forest (accuracy = 82.1%, Positive Predictive Value (PPV) = 81.4%, sensitivity = 82.1%), logistic regression (accuracy = 81.1%, PPV = 80.1%, sensitivity = 81.1%) and decision tree (accuracy = 82.1%, PPV = 81.2%, sensitivity = 82.1%). In terms of accuracy, compared to logistic regression, while random forest performed similarly, decision tree had a significantly lower discrimination ability (p-value $<0.05$) with AUCś equal to 85.0, 86.9, and 79.9, respectively.

In addition to the ECG signal, a relevant study by Anand Kumar et al. (2020) provides an intuition experimental procedure to achieve the optimum results regarding EMG classification using several machine learning models. For data analysis and interpretations, they have used a data set consisting of the sEMG signals collected from eleven subjects at five several upper limb position. The implementation of the proposed Deep Neural Network (DNN) based system proves their efficiency facing the other existing classifiers. Hence, the average accuracy obtained among the five subjects for DNN, SVM, kNN, Random Forest and Decision Tree is 98.88%, 98.66%, 90.64%, 91.78%, and 88.36% respectively.

In regard of EMG classification, a real-time automatic hand gesture recognition was proposed (Andronache et al. 2020). The suggested model using completely linked neural networks and temporal features extracted from the EMG signals. Consequently, the results of this study clearly show the accuracy of the proposed system (99.31%) compared to previous work (99.78%), whereas doubling the number of recognized gestures.

The proposed solutions ensure the function of machine learning algorithms for diseases detection. However, they only deal with a single type of bio signals. They do not take into consideration sensors that may be implanted in the leg, hand, fingers, brain or in any other body part. To overcome this limitation, we propose an accurate anomaly detection model for medical WSN based on three multivariate physiological signals. The proposal is based on an accurate data aggregation process for energy saving and reliability enhancement.

## 3 Proposed Model

Improving efficiency and reliability are fundamental to guarantee the success of the WBSNs applications. In this section, we outline and discuss the aim process of the investigated method described in Fig. 2.

As illustrated in Fig. 2, the first step consists three different vital signs from a patient, which are the Electrocardiogram (ECG), Electromyogram (EMG) and Blood Pressure (BP). In the second step, these sensed values are fused into one single signal in order to reduce the amount of used energy. Recall that the data aggregation process is based on feature extraction. In fact, the obtained data are combined from several signal attributes. When data is aggregated, groups of sampling are replaced by a summarized value, which are so close to the data source. Then, aggregated data will be transferred wirelessly to the medical expert for further diagnosis. As mentioned earlier, the focus will be on improving data reliability in Tiers 3.

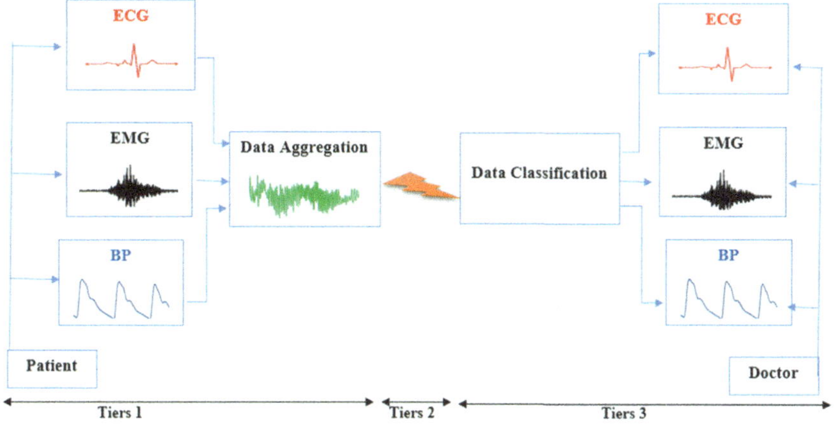


**Fig. 2** Flowchart of the proposed model

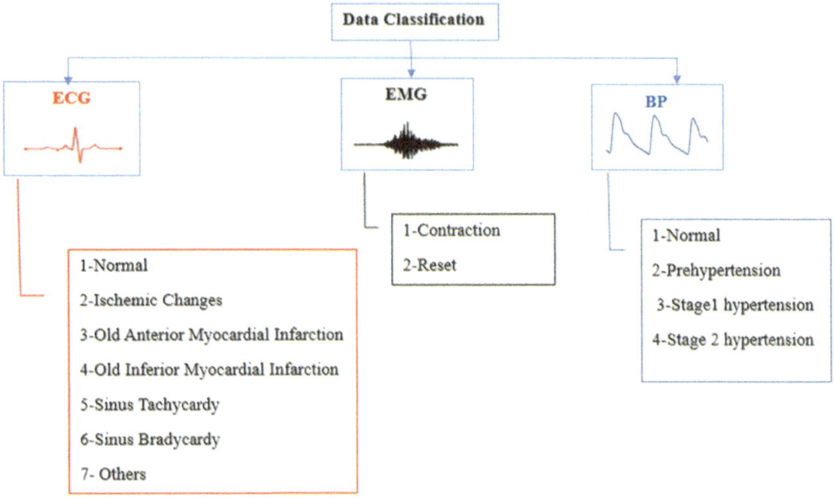


**Fig. 3** Outputs classification

**Fig. 4** Applied ML steps

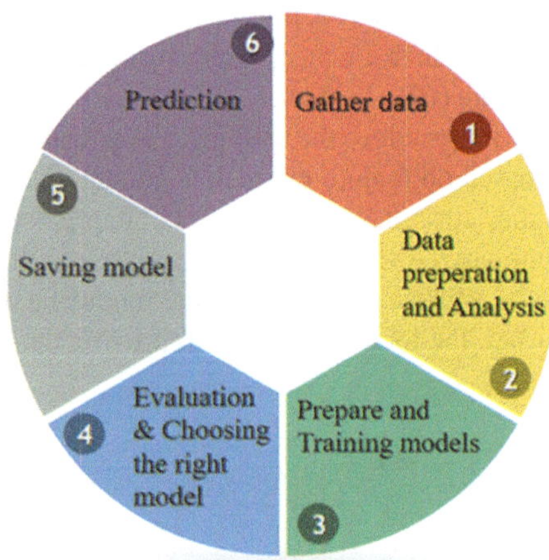


To this end, a suitable data classification model based on several supervised machine learning (ML) algorithms such as Support Vector Machine (SVM), K-Nearest Neighbors (K-NN), Random Forest (RF), Decision Tree (DT) and many other algorithms is applied. Yet, performances evaluation are validated by simulations carried out under Python tools. Finally, a comparative results based accuracy and sensitivity metrics is derived.

For the system view, it is necessary to identify the extracted diseases. To accomplish this end, Fig. 3 introduce the major existing outputs from each examined signal.

A precise process is applied to validate the simulation results. These steps are followed in order to train the studied ML algorithms. These are outlined in Fig. 4.

| | sex | age | height | weight | QRS | QRS_duration | PR_interval | QT_interval | T_interval | heartrate | ... | Weight_BP | Systolic_BP | Diastolic_BP | target_BP | stan_ |
|---|---|---|---|---|---|---|---|---|---|---|---|---|---|---|---|---|
| 0 | 1 | 56 | 165 | 64 | 25 | 81 | 174 | 401 | 149 | 53 | ... | 63 | 161 | 89 | 3 | 0.029 |
| 1 | 0 | 55 | 175 | 94 | 28 | 100 | 202 | 380 | 179 | 71 | ... | 50 | 160 | 93 | 3 | 0.036 |
| 2 | 1 | 40 | 160 | 52 | 77 | 77 | 129 | 377 | 133 | 70 | ... | 47 | 101 | 71 | 0 | 0.043 |
| 3 | 1 | 49 | 162 | 54 | 67 | 78 | 0 | 376 | 157 | 67 | ... | 65 | 123 | 73 | 1 | 0.032 |
| 4 | 0 | 44 | 168 | 56 | 61 | 84 | 118 | 354 | 160 | 64 | ... | 49 | 126 | 78 | 1 | 0.032 |
| ... | ... | ... | ... | ... | ... | ... | ... | ... | ... | ... | ... | ... | ... | ... | ... | |
| 373 | 1 | 20 | 157 | 57 | 43 | 81 | 151 | 363 | 166 | 75 | ... | 65 | 114 | 61 | 0 | 0.094 |
| 374 | 1 | 53 | 160 | 70 | -37 | 80 | 199 | 382 | 154 | 63 | ... | 61 | 131 | 75 | 1 | 0.108 |
| 375 | 0 | 36 | 166 | 68 | -85 | 108 | 176 | 365 | 194 | 84 | ... | 90 | 170 | 90 | 3 | 0.099 |
| 376 | 1 | 32 | 155 | 55 | 54 | 93 | 106 | 386 | 218 | 80 | ... | 59 | 119 | 61 | 0 | 0.091 |
| 377 | 1 | 78 | 160 | 70 | 28 | 79 | 127 | 364 | 138 | 75 | ... | 70 | 124 | 69 | 1 | 0.082 |

378 rows × 24 columns

**Fig. 5** Part of the used aggregated data set

# 4 Performance Evaluation

## 4.1 *Gathering Data*

Gathering data is an essential step in solving any supervised machine learning problem. Indeed, the text classifier can only be as good as the data set it is built from. To attain the data science goals, the world's largest data science community "Kaggle" with powerful tools and resources is used. For more clarification, Fig. 5 presents some part from the finale used data set.

As shown in this data set, the proposed data classification is based on the feature extraction process. The used attributes for the studied signal are: Age, sex, height, weight, QRS, QRS duration, PR interval, QT interval, T interval, heart rate, standard deviation, root mean square, min, max, zero crossing, systolic, and diastolic.

## 4.2 *Data Preparation and Analysis*

Before the fused data can be fed to a model, it needs to be transformed to a clear format. First, the gathered data samples may be in a specific order. Thus, it is necessary to find the pairwise correlation of all columns in the data frame. Like shown in Fig. 6, all the value in the diagonal equal to 1; that is meaning there isn't any information associated with the ordering of samples that influence the relationship between texts and labels.

The process of training an machine learning (ML) model involves providing a ML algorithm with training data to learn from. The training data must contain the correct answer, which is known as a target or target attribute. In this case, three nominated Target: $Target_H$ for ECG signal, $Target_M$ for EMG and $Traget_B$ for BP signal are used. The presented pie diagram gives the outputs classification percent for each studied signal.

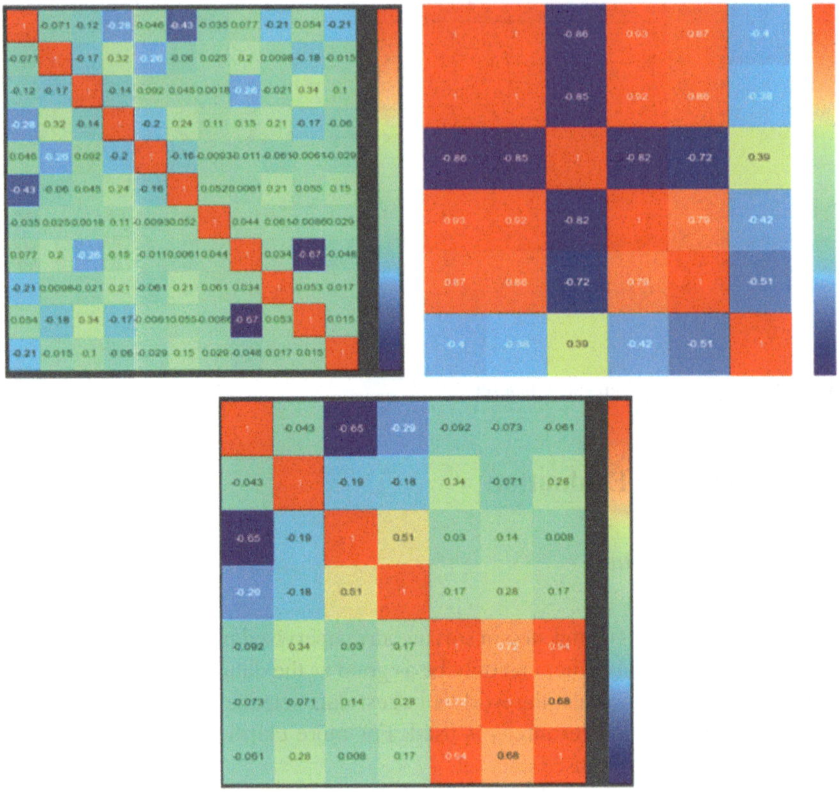

**Fig. 6** Pairwise correlation: (left) ECG, (right) EMG, (down) BP

The classification of electrocardiogram (ECG) signals plays an important role in diagnoses of heart diseases. Figure 7a presents a pie diagram of ECG classification divided into seven arrhythmia types. As shown, the highest detection probability is derived for a normal condition until 64.8%. In addition, over 11.6% of the monitored patient may suffers from ischemic changes which describe trouble to the small blood vessels in the brain. The myocardial infarction is defined as the irreversible necrosis of heart muscle producing from a decrease in blood supply to the heart due to coronary artery occlusion. For that, the myocardial infarction for both old anterior and inferior raises also a significant rate (4%) should be detected. Two others essential diseases cause irregular heart rhythms which are the tachycardia and the bradycardia. The first one is defined when a heart rate greater than 100 beats/min, and the second one when the sinus node produces an electrical charge at a slower rate than normal (60–100 beats/min). Finally, the last classification is reserved for the others diseases related to ECG signal.

The previous investigated signals result from changes in the muscle movements. For this end, the EMG classification (see Fig. 7c) attain till 84.9% of contraction

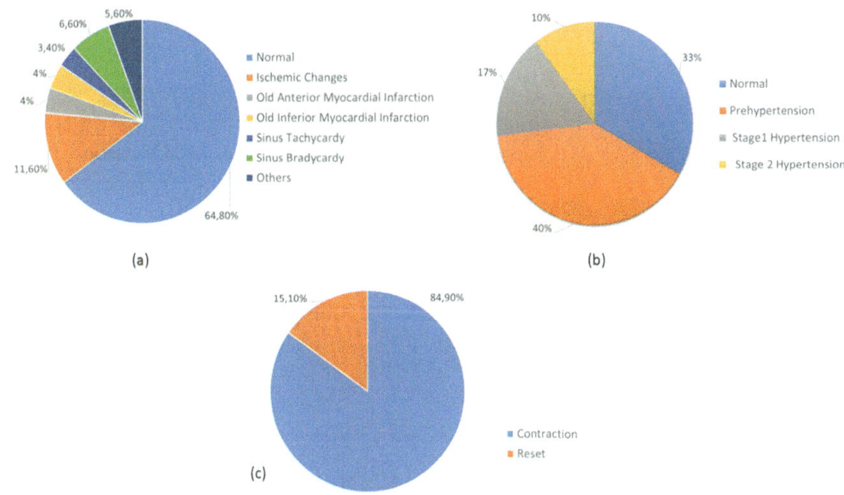


**Fig. 7** Pie diagram for: **a** ECG, **b** BP, **c** EMG

which results from the coordinated trouble of each of the muscle cells. However, merely 15.1% of training data proclaim a reset situation.

In this regard, as mentioned in the related works that the blood pressure measurements are a significant factor for cardiovascular diseases. The pie diagram for the BP also takes place. Like outlined in Fig. 7b, four significant citations are detected. Indeed, 33.3% of the studied dataset indicates a normal health status. However, a highest probability (39.7%) of a pre-hypertension situation is outlined. Two different stages of hypertension are mentioned. They are characterized by abnormally high blood pressure on the lining of the arteries. The major causes of this hypertension is under stress or during physical exertion (Fig. 7).

## 4.3 *Evaluation and Model Choice*

For the system model, it is necessary to outline the used supervised machine learning (ML) algorithms. Thus, the training model is indicated in Fig. 8 below.

Like humans, machines are able of learning in several ways. Three machine learning strategies are identified which are supervised, unsupervised, and reinforcement. In the current work, the focus will be on supervised learning algorithms, and the well-known examples are applied.

As introduced in Fig. 10, the used algorithms are: Support Vector Machine (SVM), Random Forest (RF), Decision Tree (DT), Gradient Boosting (GB), Logistic Regression (LR), Naive Bayes and the k-Nearest Neighbors (KNN).

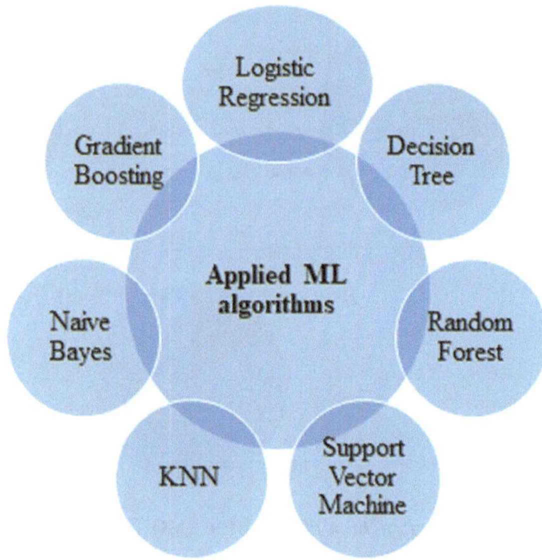


**Fig. 8** Applied ML algorithms

Firstly, the Support Vector Machine (SVM) which is greatly preferred by many applications as it produces efficient accuracy with less computation power. Secondly, the Random Forest (RF) is chosen for their flexibility. It ensures the network lifetime without hyper-parameter tuning. A tree has many analogies in real life, a Decision Tree (DT) can be used to visually and explicitly perform decision making. Besides, the Gradient Boosting (GB) is typically used with decision trees of a fixed size as base learners to enhance the optimization. Then, to assign observations for a discrete set of classes Logistic Regression (LR) algorithm is used. Finally, two simple and powerful algorithms for predictive modeling are used which are the Naive Bayes and the k-nearest neighbors (KNN).

To properly perform a model, it should always determine if it will do a good job of predicting the target on new and future data. Because future instances have unknown target values, it need to check the accuracy metric of the ML model. Then, this assessment is used as a proxy for predictive accuracy on future data. From this observation, the data set has a total of 9072 data samples and it was divided into (80%) for training and (20%) for testing. The model selection library of the Scikit Learn contains the train test split method that allows to do this separation.

To describe the performance of a classification model, it is essential to calculate the confusion matrix between actual and prediction outputs. In fact, the previous mentioned algorithms are trained and the accuracy score for each signal is calculated.

In this subsection, only the obtained confusion matrices for ECG signal are presented in (Figs. 9, 10, 11 and 12). The same process is applied for the remaining EMG and BP signals.

For more justification, the numerical results of calculated accuracy score for each EMG and BP signals are displayed in Fig. 13.

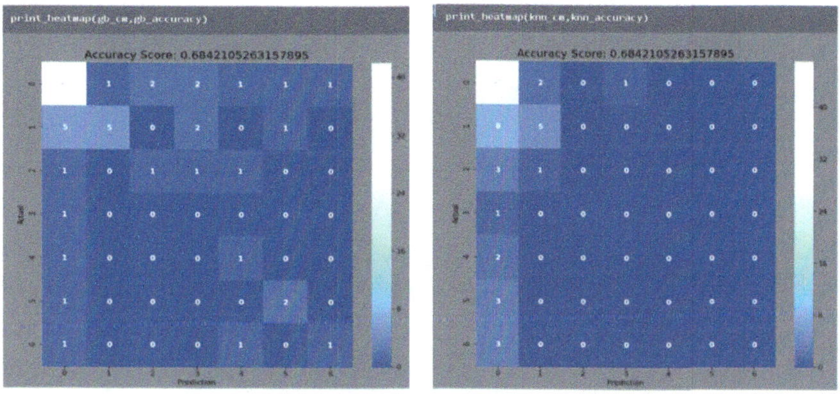


**Fig. 9** Confusion matrix: **a** Gradient boost, **b** KNN

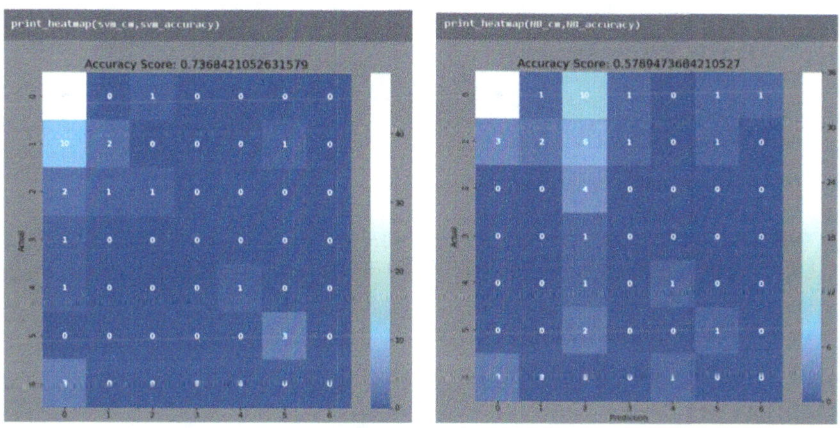


**Fig. 10** Confusion matrix: **a** SVM, **b** Naive Bayes

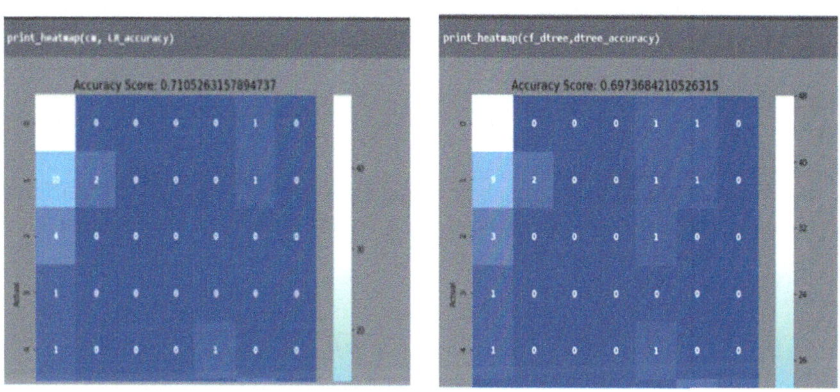


**Fig. 11** Confusion matrix: (left) Logic regression, (right) Decision tree

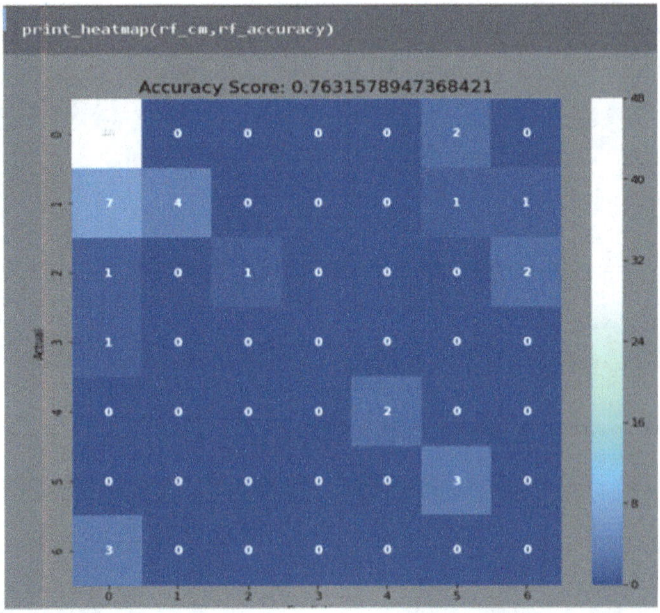


**Fig. 12** Confusion matrix based random forset

accuracy of EMG algorithms :

| | Model | accuracy |
|---|---|---|
| 4 | Random_Forest | 0.973684 |
| 6 | GradientBoost | 0.973684 |
| 0 | Nearest_Neighbors | 0.960526 |
| 3 | Decision_Tree | 0.960526 |
| 1 | Linear_SVM | 0.947368 |
| 2 | logic_regression | 0.934211 |
| 5 | Naive_Bayes | 0.934211 |

accuracy of BP algorithms :

| | Model | accuracy |
|---|---|---|
| 3 | Decision_Tree | 1.000000 |
| 4 | Random_Forest | 1.000000 |
| 6 | GradientBoost | 1.000000 |
| 1 | Linear_SVM | 0.960526 |
| 0 | Nearest_Neighbors | 0.934211 |
| 5 | Naive_Bayes | 0.907895 |
| 2 | logic_regression | 0.815789 |

**Fig. 13** Obtained accuracy: **a** EMG, **b** BP

After evaluation, to choose the best model classifier, the performance metrics in terms of accuracy and sensitivity need to be discussed. For that, the mathematical equation of accuracy score is expressed in (1):

$$Accuracy_{score}(\%) = \frac{TP}{TP + FP + TN + FN} \quad (1)$$

**Table 1** Comparative results

| Model name | ECG | EMG | BP | Sensitivity |
|---|---|---|---|---|
| Random forest | 0.7631 | 0.9736 | 1.0000 | 0.9122 |
| SVM | 0.7368 | 0.9473 | 0.9605 | 0.8859 |
| Logic regression | 0.7105 | 0.9342 | 0.8157 | 0.8596 |
| Decision tree | 0.6973 | 0.9605 | 1.0000 | 0.8859 |
| KNN | 0.6842 | 0.9605 | 0.9342 | 0.8815 |
| Gradient boost | 0.6842 | 0.9736 | 1.0000 | 0.8201 |
| Naive bayes | 0.5789 | 0.9342 | 0.9078 | 0.8069 |

Where:

- $TP$: True Positive
- $FP$: False Positive
- $TN$: True Negative
- $FN$: False Negative

For a medical diagnosing, sensitivity refers to the test's ability to correctly detect ill patients who do have the condition. Mathematically, this can be expressed in (2) as:

$$Sensitivity\ (\%) = \frac{TP}{TP + FP} \tag{2}$$

At last, the results are illustrated in Table 1. From the comparative results, it is clear that the Random Forest outperforms Support Vector Machine, Logic Regression, Decision Tree, k Nearest Neighbors, Gradient Boost and Naive Bayes in terms of accuracy (76%, 97% and 100% for ECG, EMG and BP respectively) with a sensitivity percent equal to 91.22%. Yet, the Gradient Boost algorithm raises also a high accuracy score till 97.36% especially for EMG signal. In other part, Decision Tree and Gradient Boost look as a good classifiers (until 100%) for a blood pressure movements. In addition, the SVM, DT and KNN algorithms reach a high sensitivity level over 88%. Hence, a negative result in a test with high sensitivity is useful for ruling out disease. A high sensitivity test is reliable when its result is negative, since it rarely misdiagnoses those who have the disease. Thus, this decrease proves the reliability of the three transferred aggregated medical signals based on different machine learning algorithms. By this ways, the aggregation operation does not affect the signal quality at receiving side, and the proposed approach recreates the signal/output for decision making.

Finally, the Random Forest (RF) model has the highest accuracy and sensitivity. So it is the better classifier in this case. The final applied step is to save this model using the pickle operation for future prediction.

## 5 Discussion

Decision making based machine learning algorithms have been observed in diverse fields, such as engineering (Cui et al. 2021), industry (Peters et al. 2020), physics (Han et al. 2020) or biology (Bing et al. 2020). Actually, it seems that scientists have a tight appreciation of the person risk. The reliability assessment in medical healthcare applications is unavoidable because of the limit calculation speed, memory effects and finite transmission velocity. Therefore, accuracy and sensitivity metrics have been expressed more precisely in each training model.

## 6 Conclusion

In this chapter, a process for reliability assessment of multiple health signals has been described. The system operators require several supervised machine learning algorithms to evaluate decision making at receiving side. In order to do it for multiple physiological signals, an efficient data aggregation operation to reduce the amount of used energy has been applied. By this manner, energy consumption could be reduced and the data reliability is enhanced. The required results yield the efficiency of the Random Forest algorithm in terms of accuracy (97%) and sensitivity (92%). Furthermore, a future direction of research is the real implementation of the Random Forest classifier and how to enhance the security communication in WBSN.

**Acknowledgements** This research framework is within a collaboration between the National School of Engineering of Gabes, the Digital Research Center of Sfax (CRNS), Tunisia and Chemnitz University of Technology, Germany. The authors draw up many thanks the European exchange program for supporting their work.

## References

Alghamdi, A.S., Polat, K., Alghoson, A., Alshdadi, A.A., & Abd El-Latif, A. (2020) A novel blood pressure estimation method based on the classification of oscillometric waveforms using machine-learning methods. *Applied Acoustics*, *164*, 107279.

Alkaabi, L., Ahmed, L., Al Attiyah, M., & Abdel-Rahman, M.(2020). Predicting hypertension using machine learning: Findings from Qatar biobank study. *LOS One*, *15*(10).

Altamimi, A. S. H., Al-Dulaimi, O. R. K., Mahawish, A., Hashim, M. M., & Taha, M. S. (2020). Power minimization of WBSN using adaptive routing protocol. *Indonesian Journal of Electrical Engineering and Computer Science*, *19*(2), 837–846. ISSN: 2502-4752.

Anand Kumar, M., & Suman, S. (2020). An experimental study on upper limb position invariant EMG signal classification based on deep neural network. *Biomedical Signal Processing and Control*, *55*.

Andronache, C., Negru, M., Neacsu, A., Cioroiu, G., Radoi, A., & Burileanu, C. (2020). Towards extending real-time EMG-based gesture recognition system. In *43rd International Conference on Telecommunications and Signal Processing (TSP)*, Milan, Italy (pp. 301–304).

Bahae, A., Abdelillah, J., & El Haziti, M. (2020). Wireless body area networks: A comprehensive survey. *Journal of Medical Engineering Technology, 44*(3).

Belhaj Mohamed, M., Meddeb-Makhlouf, A., & Fakhfakh, A. (2018). Correlation for efficient anomaly detection in medical environment. In *International Wireless Communications and Mobile Computing Conference (IWCMC)*.

Bing, L., Yinzi, L., Yu, W., David, I. W., & Brent, R.Y. (2020). Application of mechanistic modelling and machine learning for cream cheese fermentation pH prediction.*Journal of chemical Technology and Biotechnology*.

Cui, L. J., Sun, M. Y., Cao, Y. I., et al. (2021). A novel tolerance geometric method based on machine learning. *Journal of Intelligent Manufacturing*.

Dehkordi, S. A., Farajzadeh, K., Rezazadeh, J., et al. (2020). A survey on data aggregation techniques in IoT sensor networks. *Wireless Networks, 26*, 1243–1263.

Han, W., Hua, B., & Xiulin, R. (2020). Machine learning prediction of thermal transport in porous media with physics-based descriptors.*International Journal of Heat and Mass Transfer, 160*, 120176. ISSN 0017-9310.

Kunal Kishan, S., & Mayank, D. (2020). Privacy preserving data aggregation in wireless body sensor network. In *Proceedings of the 2nd International Conference on IoT, Social, Mobile, Analytics Cloud in Computational Vision Bio-Engineering (ISMAC-CVB)* (13 P.).

Mehrani, M., Attarzadeh, I., & Hosseinzadeh, M. (2020). Optimum data collection and fusion schemes in WBSN. *International Journal of Sensor Networks*.

Mohamed, M. B, Meddeb-Makhlouf, A., & Fakhfakh, A. (2019). Intrusion cancellation for anomaly detection in healthcare applications. In *International Wireless Communications and Mobile Computing Conference (IWCMC)*.

Mohamed, M. B, Meddeb-Makhlouf, A., Fakhfakh, A., & Kanoun, O. (2020). Intrusion detection based on correlation of multiple health signals in WBSN. In *20th International Multi-Conference on Systems, Signals Devices (SSD)*.

Peters, P., Klieštik,T., Musa,H., & Durana, P. (2020). Product decision-making information systems, real-time big data analytics, and deep learning-enabled smart process planning in sustainable industry 4.0. *Journal of Self-Governance and Management Economics*, 16–22.

Plawiak, P., & Abdar, M. (2020). Novel methodology for cardiac arrhythmias classification based on long-duration ECG signal fragments analysis. *Biomedical Signal Processing, Series in Bio-Engineering*.

Tarique, R., Sunil, K., Akshay, V., Prateek R. G., & Arvind, K. (2020). Co REERP: Cooperative reliable and energy efficient routing protocol for intra body sensor network (intra WBSN). *Wireless Personal Communications*.

Verma, A.R., Gupta, B., & Bhandari. C. (2020). A comparative study of ECG beats variability classification based on different machine learning algorithms. *Augmented Human Research*.

# Accelerated Moving Humans Detection Algorithm using Combined Global Descriptors on GPU Based on CUDA

**Haythem Bahri, Marwa Chouchene, Randa Khemiri, Fatma Ezahra Sayadi, and Mohamed Atri**

**Abstract** Day by day, the ability to detect and to identify automatically the objects among images and videos without constraint has becoming more and more important. The security systems, robots, smartphones and smart devices need to know the semantic meaning of image. The increase in object detection and identification algorithms is essentially related to the increase in complex object specification and authentication techniques. This could be resolved only when using the parallel architectures that can support heavy parallel processing such as GPU. In this chapter, we propose to present an implementation of moving humans detection algorithm on GPU based on the programming language CUDA. We proposed an implementation of an algorithm to extract the image features using the Fourier descriptor on GPU. We have proposed a second implementation to extract the image features based on the HOG descriptor on GPU. To detect the moving objects, we have implemented a background subtraction algorithm based on the GMM: Gaussian Mixture Model on GPU. In order to integrate these implementations in the main moving humans detection algorithm, the use of preprocessing and filtering techniques is necessary at this level as well as the CCL: Connected Component Labeling method which allows

H. Bahri (✉) · M. Chouchene · R. Khemiri · F. E. Sayadi
Laboratory of Electronics and Microelectronics, Faculty of Sciences of Monastir, University of Monastir, Monastir, Tunisia
e-mail: bahri.haythem@hotmail.com

M. Chouchene
e-mail: ch.marwa.84@gmail.com

R. Khemiri
e-mail: randa.khemiri@gmail.com

F. E. Sayadi
e-mail: sayadi_fatma@yahoo.fr

M. Atri
College of Computer Science, King Khalid University, Abha, Saudi Arabia
e-mail: Matri@kku.edu.sa

 
O. Kanoun and N. Derbel (eds.), *Advanced Sensors for Biomedical Applications*, Smart Sensors, Measurement and Instrumentation 38,
https://doi.org/10.1007/978-3-030-71225-9_5

extracting the Moving objects from the rest of the image. The implementation of such kind of algorithm on GPU allows a great performance in terms of execution time.

**Keywords** Human detection · Fourier descriptor · Compute Unified Device Architecture · Histogram of oriented gradients · Support Vector Machine · Graphics Processing Unit · Gaussian Mixture Model

## 1 Introduction

The past decade has been marked by tragic events of an accidental or intentional nature such as terrorist attacks. This has made companies more demanding in terms of safety and prevention. As a result, digital video surveillance systems are now being developed and installed in residential, public and work environments. This technology has gradually become an essential means of contributing to the improvement of security in cities.

The theory of video surveillance and its applications have grown considerably in recent years with great results. However, they lack processing efficiency, speed of execution and low energy consumption. These realities are serious drawbacks for real-time applications, such as real-time people detection systems. There are different ways that video surveillance algorithms analyze data, but they are almost all cumbersome and complex in terms of calculation. This increases the CPU power consumption and runtime. Indeed, these systems are generally subject to execution time constraints, typically real time. Unexpectedly, the treatments to be performed by these applications are increasingly complex for several reasons, including: the consideration of more sophisticated algorithms; increasing the resolution of the images processed as well as that of their bit rate in a video acquisition; the complexification of the nature of the targeted models; to quote only those.

As a result, meeting the time constraints becomes a more difficult challenge. One way to overcome these problems is to wait for the video surveillance algorithms to find new ideas for data processing or to try to implement heavy parts of an algorithm on hardware such as FPGAs and GPUs. By taking advantage of the nature of the hardware for parallel computing, the processing time can be significantly reduced while maintaining low power for complex calculations. Specifically for GPUs, which have become increasingly programmable over the last decades thanks to NVIDIA the parallel computing leader with its intuitive CUDA or Compute Unified Device software Architecture and its highly optimized GPGPU hardware. Data can be transferred into their vast memory and using billions of virtual threads, great results can be obtained by extracting the inherent parallelism of a video surveillance algorithm. All this boils down to better performance and realistic use of video surveillance programs in everyday life.

The multi-core architectures and GPUs as graphics processing units are being increasingly used in modern video surveillance systems because they provide energy and cost efficient platforms. To meet the real-time processing, we harness those kinds of architectures by extracting data level parallelism in such algorithms. Video human detection is a crucial step in many visual computing applications like human detection in video surveillance (Chen et al. 2014; Kumar et al. 2016), driving assistance (Pedersoli et al. 2014; Campmany et al. 2016), roadside unit for intersection assistance (Bauer et al. 2010; Weimer et al. 2011) and object movement verification and tracking (Sankaranarayanan et al. 2008). Most of them have high computational loads, storage, and bandwidth requirements. Hence the idea to implement that kind of video surveillance algorithms like moving humans detection on GPUs and other multi-core architectures.

In this paper, a parallel implementation of a human movement detection algorithm on the GPU architecture is discussed in detail. At first, the paper focuses on some algorithms such as Gaussian Mixture Model (GMM) to extract the foreground from video on GPU. Secondly, some operating ran on the CPU such as the morphological operator for noise removal in a video frame and Connected Component Labeling (CCL) for identifying the foreground of moving object. Then two algorithms are implemented on the GPU to extract the features from the moving object: Fourier descriptor and Histogram of Oriented Gradients (HOG) descriptor. Finally, the adopted classifier Support Vector Machine (SVM) uses the feature descriptors to provide the classification result humans from non-humans moving objects. The GPU algorithms were tested by executing different frame sizes on Geforce GTX480 GPU.

The structure of the paper is as follows: in Sect. 2, we explain the used methods in the proposed moving humans detection algorithm. In Sect. 3, we detail the main structure of our proposed algorithm and its implementation on GPU. Section 4 is devoted to experimental results and discussions of the implementation of the proposed algorithm. In Sect. 5, we conclude our paper and we present future work.

## 2 Methods and Algorithms

The main goal of the proposed algorithms is to extract the image regions corresponding to moving object from the rest of the video frame and then classify this object in order to detect the human using SVM. Therefore, we use Gaussian Mixture Model the most efficient background subtraction model to extract the foreground. After that, we use some preprocessing to eliminate noisy areas in the foreground map as blur and threshold filter. Next, we apply the morphological operator to provide more compact regions without editing the original forms. Then we extract the Fourier and the HOG descriptor for each moving object which will be an input of the adopted classifier SVM. All steps of the global algorithm will be detailed in the following subsections.

## 2.1 Gaussian Mixture Model

The GMM approach was initially proposed by Friedman and Russell in Friedman and Russell (1997) to subtract the background. Later, many efficient updates of GMM equations are given. Indeed, the GMM update equations are continually enhanced to simultaneously select the appropriate number of components for each pixel, allowing the model to not fully adapt to the scene but also reduces processing time and improves the segmentation. Until the model of Stenger et al. (2001) that suggests the number of components of a hidden Markov model in an off-line training procedure. The GMM stores M separated normal distributions for each pixel (parametrized by mean $\mu_i$, variance $\sigma_i^2$, mixing weight $\omega_i$, where i = 1, 2…M) and M typically between 3 and 5 (depending on the complexity of the scene). The probability distributions for pixel value $I_t$ is expressed as follows:

$$P(I_t) = \sum_{i=1}^{M} \omega_i N(I_t; \mu_i, \sigma_i^2 I) \tag{1}$$

where $P(I_t)$ is the pixel value probability, it is shown for 1D pixel value $I_t \in \{0, 1 \ldots 255\}$, M = 3, $\omega_i = \{0.3, 0.3, 0.4\}$, $\mu_i = \{70, 120, 200\}$, $\sigma_i^2 = \{25, 30, 10\}$. In this method, a foreign object appearing in the scene will be represented by some additional components with low weights and high variances. Therefore, we can conclude that the background-foreground segmentation can be achieved by selecting the mixture of the N components of the highest weight to variance ratios as a background model, and the remainder as foreground model. Due to acceptance criterion, the GMM contains the intrinsic assumption that background pixels have low variance.

The blend model contains both the model distributions of the background and those of the foreground model. That is why the minimum logical value for the number of distributions is 3, so that 2 of them can be assigned to handle the bimodal background of the scene and dedicate a distribution to describe the foreground. Once the current state of $i$ is estimated, a threshold must be defined to separate the model's background distributions from those belonging to the foreground model. Distributions likely to be part of the background are those with a high weight, and with low variances.

To combine these two factors for each pixel, all existing distributions are classified by a criterion of $\omega_i/\sigma_i$. This factor reaches its peak when $\omega_i$ is large and inversely $\sigma_i$ small. Therefore, the most likely mixture components that can model the background are those with small standard deviations (intensities do not vary much) and high probability of occurrence. After the distributions are ranked on the basis of the factor $\omega_i/\sigma_i$, the weights of the corresponding distributions are summed and the result is checked with a predefined threshold:

$$B = argmin_b \left( \sum_{i=1}^{M} \omega_i > T \right) \tag{2}$$

$B$ is the minimum number of distributions belonging to the background among the $M$ distributions available for each pixel. $T$ is the fraction of the overall weight attributed to the model of the background. A low value of $T$ corresponds to the unimodal case, usually a single distribution will correspond to the background. A high value corresponds to the multimodal case, several distributions will be included in the background. A multimodal distribution is often caused by a repetitive background motion such as the movement of leaves on a tree or flag. $T$ is a value between 0 and 1. This threshold is a measure of the minimum part of the data that should be taken into account by the merits. Thus, the weight of the most likely background distributions are accumulated until the sum exceeds the threshold $T$. $B$ is then the number of these distributions.

## 2.2 *Fourier Descriptor*

Fourier descriptor is one of the tools that represent all image information's in a coefficients vector. This model uses Fourier Fast Transform (FFT) to extract different features of an image. Those features were represented in coefficients vector that will be used as an input for learning machine to identify their class (SVM, neuron network, etc...). The descriptor includes all the information of the object in the image and keeps the invariance characters when applying the geometric, noise and lighting transformations. It can be in a global model to represent all pixels in the image or in a local one where it shows a region of interest or interesting points. We can cite some models of descriptor such as a color descriptors, shape descriptors and texture descriptors. We are interested in shape descriptors model as an application domain. In this context, we studied the Fourier descriptor for the images recognition and classification. It uses the FFT to extract all information of a color image. The data represented in a vector will be used as input to the classifier in order to identify its class. Fourier descriptors can be used in various applications such as object recognition and forms, tracking of objects or person and real-time image recovery. We can describe the Fourier descriptor steps as follows: initially, read the image then transform these plans using FFT, afterward, group the FFT images blocks and calculate the square modules of these resulting images and finally, apply the suitable algorithm to determine the Fourier descriptor (Bahri et al. 2017).

## 2.3 *Histogram of Oriented Gradients*

The computation of the HOG is based on evaluating normalized local histograms of image gradient direction in a dense grid. This method tries to characterize local object appearance and shape by the distribution of local intensity gradient directions. The main advantage of the HOG representation is the fact that it is robust to illumination variance because gradient directions of local regions do not change with illumination variations. It is also robust to small deformations because slight shifting and affine deformations make small histogram value changes.

## *2.4 Preprocessing: Blur Filter, Thresholding and Opening Morphological Operator*

In our algorithm, we used two pretreatment tools in order to destroy the image noise by the effect of background subtraction. The blur filter for smoothing the image and eliminate noises and details and thresholding to differentiate the relevant pixels from the rest. In the following we define these two processes to better understand their choice in our contribution.

### 2.4.1 Blur Filter

The blur filter or also the homogeneous filter is the simplest method of smoothing an image. You just have to calculate the neighborhood average of a pixel and give that value. You must choose the right size of the smoothing matrix. If it is too big, the small features of the image may disappear and the image looks blurry. If it's too small, you can not eliminate the noise from the image. Each channel of the input image is processed independently. It is smooth by sliding its smoothing matrix on the image. Each pixel value will be calculated based on the value of the array and the overlapping pixel value of the original image. In mathematical terms, we do a convolution operation on an image with a smoothing matrix. The matrix we apply to an image makes the difference with the result of smoothing. What we do for this filter affects the average values of the neighbors of a pixel.

### 2.4.2 Thresholding

Thresholding is the simplest segmentation method, it consists in separating the regions of an image corresponding to objects that we want to analyze. This separation is based on the variation of intensity between the pixels of the object and the pixels of background. To differentiate the pixels that interest us from the rest that will ultimately be rejected, we compare the intensity of each pixel with a threshold. Once the important pixels are separated correctly, we can set them with a determined value to identify them. That is, we can assign them a value of 0 corresponding to black or 255 corresponding to white or any other value that suits the needs. All thresholding algorithms take a source image and a threshold value as input and produce an output image by comparing the source pixel value with the threshold. If the source pixel is greater than the threshold, then the pixel of the output image receives a certain value. Otherwise, if the source pixel is below the threshold, the output pixel receives another value different from the first one.

### 2.4.3 Opening Morphological Operator

Mathematical morphology has rapidly become, since its introduction in the 1960s, a fundamental theory of image processing and analysis. The operators it offers can provide tools for the entire image processing chain, from pre-processing (filtering, contrast enhancement) to segmentation and scene interpretation. One of the important features of these operators is that they are non-linear. They make it possible to transform the images, to extract characteristics, objects or even measurements by an analysis associating properties of the objects themselves (shape, size, appearance ...) and properties of the context (local neighborhood or relations with other objects). In the following, we introduce the four basic operations of mathematical morphology: erosion, dilation, opening and closing. Some immediate applications of these operations will be illustrated.

## 2.5 *CCL: Connected Component Labeling*

Connected Component Labeling (CCL) is almost the basis of all algorithmic chains in image processing when it comes to analyzing regions in a scene. More generally, it is present when it is necessary to group related pixels. We also find it to label segments, to detect contours, or to find optical character recognition ORC or Optical Character Recognition. Standard algorithms use either the pixel approach or an approach based on RLC or Run Length Coding. These two approaches use a large number of tests, which is translated at the processor level by a blockage of the pipeline and therefore a waste of time, because the result of a test is unpredictable. Finally and this is not the least of its flaws, the pixel approach produced a lot of useless labels. Associated component labeling is a computer vision operation that extracts the connectivity information from a previously segmented image. This technique makes it possible to go from the analysis level linked to the pixel scale to a level of analysis linked to the information of the different regions of the image. It is a method widely used by convex hull extraction, segmentation, hysteresis filtering or geodetic reconstruction algorithms.

**Fig. 1** **a**-Bounding image, **b**-External bounding image

## 3 Moving Humans Detection Algorithm and Its Implementation on GPU

After the brief overview of the different methods used, we try to present the overall context of the Moving humans detection algorithm based on these techniques. The GMM presents the first and the main step to extract the foreground image using the background subtraction implemented on GPU using CUDA (Pham et al. 2010). Then, we applied on CPU at first the blur filter to reduce noise and detail image and secondly the thresholds filter to eliminate noisy areas in the binary image, any blob smaller than the threshold size is removed. Besides this refinement, we applied the morphological operator as opening (erosion followed by dilation) or closing (dilation followed by erosion) to clean-up spurious responses to detach touching objects and fill in holes for single objects. Therefore, we performed the opening operator to the binary image from the GMM on the CPU to provide more compact regions without editing the original shapes. To find the relevant moving objects, we used the Connected Component Labeling (CCL) model on CPU to extract both images: the first is the external bounding image and the second is the bounding image respectively from the binary and the original frame. In our algorithm, we have succeeded in implementing on GPU the Fourier descriptor for the external bounding image and the Histogram of Orient Gradient for the bounding image as shown in Fig. 1.

The proposed algorithm that computes the Fourier descriptor of the external bounding image is composed of three major stages. In the first stage, we resized

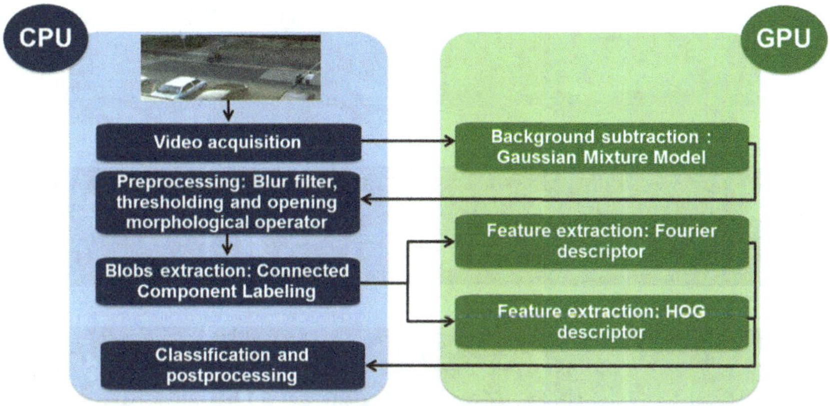


**Fig. 2** Implementation of moving humans detection algorithm

the external bounding image to 64 × 128. In the next stage, the FFT is computed and shifted to extract the feature vector. Finally, the resulting vector is normalized.

To compute the HOG, the original image inside the bounding image is also resized to 64 × 128. Then it is divided into 128 cells of size 8 × 8. We compute the HOG direction for each cell and each histogram that is quantified into 9 bins for the gradient direction range of $-90°$ to $90°$. It is necessary to work towards normalizing the local contrast of the gradient to reduce the effect of the image illumination variations. It is achieved by spatially grouping cells into overlapping blocks.

One block consists of 2 × 2 cells and any two neighboring blocks have two cells in common due to the overlapping. All the histograms of the block are concatenated into a feature vector of dimension 36. Since the bounding image consists of $15 \times 7 = 105$ blocks, the content of this image is represented by a $36 \times 105 = 3780$ feature vectors. After the extraction of the feature vectors from the external bounding and the bounding image, we used SVM to determine whether the moving object is human or not. Figure 2 summarizes the multi-stage algorithm for video human detection.

Figure 3 shows a sample output of our Moving humans detection algorithms on a pedestrian video sequence. The first row shows selected frames from the original video sequence. The second row presents the extraction of moving object using GMM implementation on GPU. The third row processes the GMM output by blur and threshold filtering respectively and more especially the morphological operators processing. The last row shows the output after the prediction with SVM displaying the bounding box and the contours around the detected human.

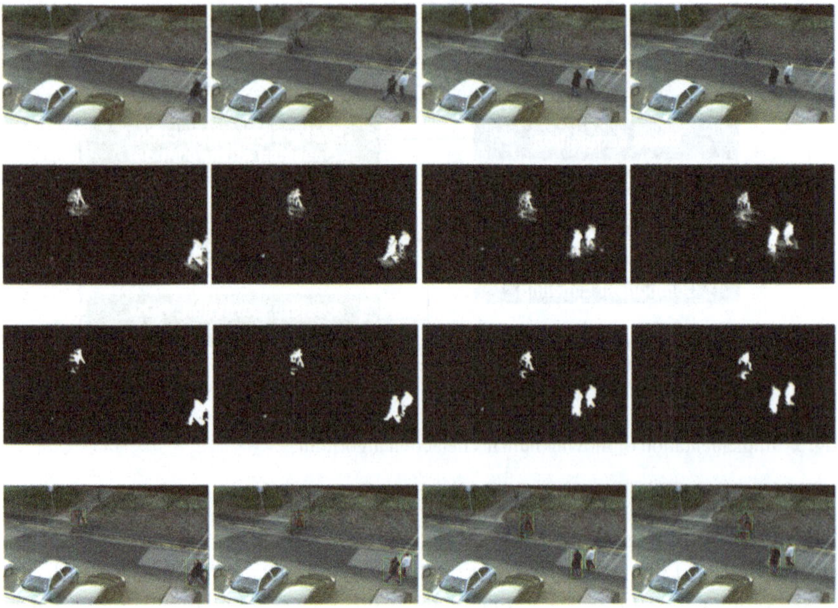

**Fig. 3** Sample output of our video surveillance algorithms of moving humans detection implementation on GPU

# 4 Experimental Results and Discussion

The experiments were carried out on a NVIDIA GeForce GTX480 graphic card and an Intel Core I7-3770 3.40 GHz CPU with 8 GB memory. The GPU software was coded in the programming language of CUDA Toolkits version 4.0. The CPU software was created in Visual C++ 2010.

## *4.1 GMM Implementation on GPU*

Figure 4 compares the performance of a sequential and a parallel implementation of GMM algorithm on different frame sizes. The result shows a significant speedup for parallel GPU implementation going up to 14.7$\times$ when compared to the sequential execution on CPU. Apart from the image data, GMM stores the background model information of mean ($\mu$), variance ($\sigma$) and weight ($\omega$) values in the memory which need to be transferred back and forth from host to device memory for successive frame computation.

We try to overlap the communication with computation time and, hence, we are able to obtain significant speedup values. It can be observed that the speedup increases with image sizes because of the increase in the total number of CUDA

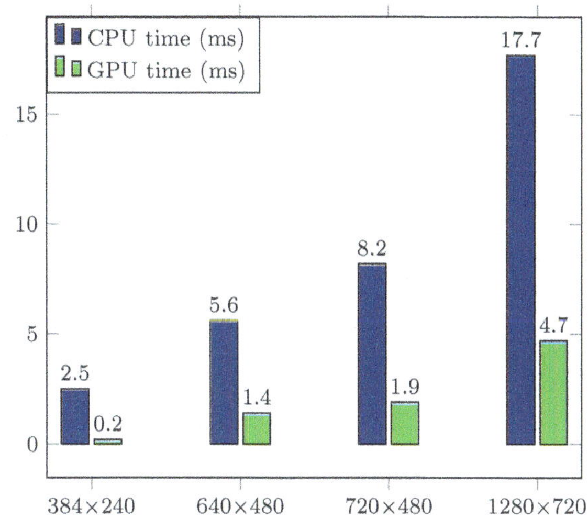


**Fig. 4** Comparison of the GMM implementation time on CPU versus on GPU

threads (each pixel is operated upon by one CUDA thread) which keeps the cores busy. For optimization, we used shared memory for storing various Gaussian mixture values thus reducing the memory access but in counterpart decreases the occupancy.

## *4.2 Fourier Descriptor Implementation on GPU*

Figure 5 compares the performance of a sequential and a parallel implementation of Fourier descriptor algorithm for different frame sizes. As shown in Fig. 5, we are able to get enormous speedup for the parallel GPU implementation going up to 5.4× when compared to the sequential execution on CPU.

This important speedup is due to several factors. First of all, inputs of this algorithm are binary values for each pixel which need very less memory for storing the entire frame and thus the transfer time is reduced to the minimal. We could also get maximum occupancy using CUDA Occupancy Calculator a spreadsheet able to determine how many blocks of threads may run in parallel on a specific GPU, depending on the number of resources (registers and shared memory) the threads use.

## *4.3 HOG Implementation on GPU*

Figure 6 compares the performance of a sequential and a parallel implementation of HOG algorithm using different frame sizes. Although HOG algorithm is hard to

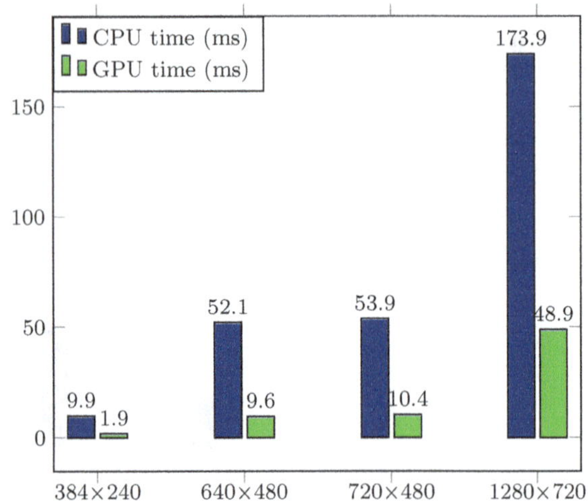


**Fig. 5** Comparison of the Fourier descriptor implementation time on CPU versus on GPU

parallelize due to many dependencies, it was possible to obtain speedup up to 44.5× for parallel GPU implementations, especially on small image sizes.

As shown in Fig. 6, the HOG implementation on GPU gave better results in terms of execution time. In fact, while a GPU gives an execution time around a few milliseconds, that on CPU exceeds 13.59 s. The overall execution time of HOG algorithm affects directly the main time of the Moving Humans detection algorithm where it resides the main process to extract all minutely details of the image.

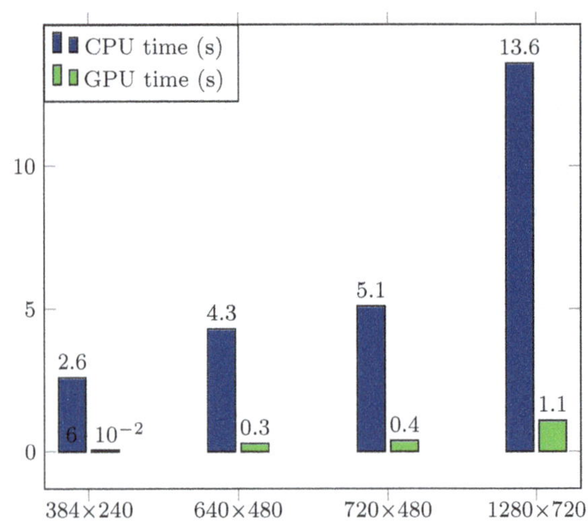


**Fig. 6** Comparison of the HOG implementation time on CPU versus on GPU

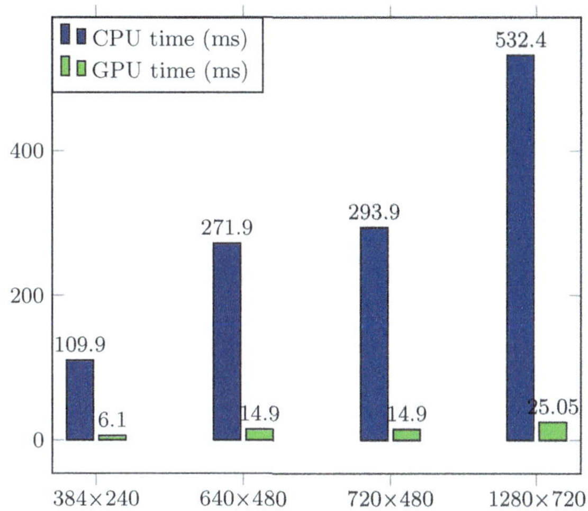


**Fig. 7** Comparison of the GMM, Fourier descriptor and HOG implementation time on CPU versus on GPU

### *4.4 Moving Humans Detection Algorithm Implementation on GPU*

To more explain the performance of our Moving Humans detection algorithm, we tried in Fig. 7 to show the total execution time of all kernels running on GPU and on CPU. These kernels serve to compute the GMM of the input video in order to extract the moving object and to compute the features of moving object using Fourier and HOG descriptors. The gain in execution time is more than 18 times at least and it is more than 20 times at most. We can improve these results by using the GPU specific memories such as constant, shared and texture memory.

The transfer time between the CPU and the GPU is not considerable but it should not be neglected. Since the size of frame increase, the transfer time is proportionally increases.

## 5 Conclusion

We have presented a hybrid approach for human detection in video surveillance systems using several techniques. The proposed algorithm extracts the foreground from the video surveillance to track the moving object. To achieve accuracy, we combined texture and form based representations to capture the human. We integrated two kinds of feature descriptor Fourier descriptor and histogram of Orient Gradient in SVM. Those techniques are implemented on GPU with successful for fast and accurate human detection. In experimental result, we compared the performance of CPU and parallel GPU implementations of GMM, Fourier descriptor, HOG and the entire detection chain. We show that the GPU parallel implementation achieved significant speedups up to a factor of 14.7× for GMM, 5.4× for Fourier descriptor, 44.5× for HOG, and up to 20× for the set of these kernels in the algorithm in comparison to CPU.

In our future work, we will implement the morphological operation and CCL in parallel on the GPU, we will focus on the accuracy of the human detection since it is one of the most important factors in video surveillance and we will explore our results to reach the real time for each resolution video.

## References


Bahri, H., Sayadi, F., Khemiri, R., Chouchene, M., & Atri, M. (2017). Image feature extraction algorithm based on cuda architecture: Case study gfd and gcfd. *IET Computers & Digital Techniques, 11*(4), 125–132.

Bauer, S., Köhler, S., Doll, K., & Brunsmann, U. (2010). FPGA-GPU architecture for kernel SVM pedestrian detection. In *In IEEE Conference on Computer Vision and Pattern Recognition Workshops (CVPRW)* (pp. 61–68).

Campmany, V., Silva, S., Espinosa, A., Moure, J. C., Vázquez, D., & López, A. M. (2016). Gpu-based pedestrian detection for autonomous driving. *Procedia Computer Science, 80*, 2377–2381.

Chen, L.-H., Wang, L.-Y., & Su, C.-W. (2014). Human detection in surveillance video. *International Journal of Pattern Recognition and Artificial Intelligence, 28*(2), 1455003.

Friedman, N., & Russell, S. (1997). Image segmentation in video sequences: A probabilistic approach. In *13th Conference on Uncertainty in Artificial Intelligence* (pp. 175–181). Burlington: Morgan Kaufmann Publishers Inc.

Kumar, P., Singhal, A., Mehta, S., & Mittal, A. (2016). Real-time moving object detection algorithm on high-resolution videos using gpus. *Journal of Real-Time Image Processing, 11*(1), 93–109.

Pedersoli, M., Gonzàlez, J., Hu, X., & Roca, X. (2014). Toward real-time pedestrian detection based on a deformable template model. *IEEE Transactions on Intelligent Transportation Systems, 15*(1), 355–364.

Pham, V., Vo, P., Hung, V. T. et al. (2010). Gpu implementation of extended gaussian mixture model for background subtraction. In *IEEE International Conference on Computing and Communication Technologies, Research, Innovation, and Vision for the Future (RIVF)* (pp. 1–4).

Sankaranarayanan, A. C., Veeraraghavan, A., & Chellappa, R. (2008). Object detection, tracking and recognition for multiple smart cameras. *Proceedings of the IEEE, 96*(10), 1606–1624.

Stenger, B., Ramesh, V., Paragios, N., Coetzee, F., & Buhmann, J. M. (2001). Topology free hidden markov models: Application to background modeling. In *8th IEEE International Conference on Computer Vision, ICCV* (Vol. 1, pp. 294–301).

Weimer, D., Köhler, S., Hellert, C., Doll, K., Brunsmann, U., & Krzikalla, R. (2011). GPU architecture for stationary multisensor pedestrian detection at smart intersections. In *IEEE Intelligent Vehicles Symposium (IV)* (pp. 89–94).

# Human Breathing Monitoring by Graphene Oxide Based Sensors

**Ammar Al-Hamry, Enza Panzardi, Marco Mugnaini, and Olfa Kanoun**

**Abstract** Non-invasive monitoring of human health is of a high importance for early detection of illnesses and improving life quality. Breath monitoring is important for detection of severe diseases such as lung cancer or sleep apnea. In this work, we introduce a breath sensor based on a graphene oxide film deposited on silver interdigitated electrode and a flexible substrate. The graphene oxide film was then thermally annealed to partially reduce the graphene oxide. The measurements of sensor impedance carried out at different humidity levels show a high decrease by several orders of magnitudes by increasing the relative humidity. Sensitivity to humid air results from the high hydrophilicity of the graphene oxide due to its oxygen functional groups. The change of transport mechanism from Nyquist plot shows the change of the sensor impedance from the capacitive behavior to a semicircle of parallel resistance and capacitance. The sensors show an ultrahigh sensitivity to humidity at high humidity values, a very low response time of less than one second and an excellent repeatability of the measurements. For tracking human breathing, the reaction on natural breathing was acquired by a digital oscilloscope together with an IoT mobile application to visualize the results in real time and store them for further processing. The sensor performance shows that it is suitable as a noninvasive and flexible breath-monitoring sensor system. The proposed sensor can be a step to flexible and cheap wearable sensors for detection of human breath and hazardous breath airborne such as COVID-19.

A. Al-Hamry (✉) · O. Kanoun
Chair for Measurement and Sensor Technology, Technische Universität Chemnitz, Chemnitz, Germany
e-mail: ammar.al-hamry@etit.tu-chemnitz.de

O. Kanoun
e-mail: olfa.kanoun@etit.tu-chemnitz.de

E. Panzardi · M. Mugnaini
Information Engineering and Mathematics, Università di Siena, Via Roma, Siena, Italy
e-mail: panzardi@diism.unisi.it

M. Mugnaini
e-mail: mugnaini@diism.unisi.it

 
O. Kanoun and N. Derbel (eds.), *Advanced Sensors for Biomedical Applications*, Smart Sensors, Measurement and Instrumentation 38,
https://doi.org/10.1007/978-3-030-71225-9_6

**Keywords** Graphene Oxide (GO) · Impedance spectroscopy · Breath monitoring · Humidity sensor · IoT

## 1 Introduction

Nowadays, IoT applications in medical care are of a high importance as they offer continuous monitoring of data to enable remote and long term diagnosis. In this context, sensors for non-invasive monitoring are considered as an important way for human health check such as breath diagnosis (Kanoun et al. 2018). The analysis of breath as an non-invasive method provides opportunities for detection of diseases and represents an alternative approach of blood analysis (Econsultancy 2020; Kim et al. 2015). On the one hand, there are the exhaled breath sensors through which several volatile organic compounds can be detected and studied to give clue to certain diseases (Lau et al. 2017). As examples, Si-doped epsilon-WO3 nanostructures (Righettoni et al. 2012), SnO2-reduced graphene oxide (Zhang et al. 2015) and nanostructured Pt-doped SnO2 sensors (Van Den Broek et al. 2018) were used to detect acetone in the breath of diabetes patients.

On the other hand, other sensors are used to monitor breath dynamics connected to diseases like sleep apnea. Several sensors were proposed, which utilize thermal sensors, reduced graphene oxide sensor or carbon nanotubes-based sensors. Caccami et al. (2018) developed wearable radiofrequency identification (RFID) integrated in face mask based on graphene oxide sensors, which show low sensitivity and low response time. A Graphite/silver nanoparticles-based sensor deposited on a paper substrate was also characterized by different tests of humidity and breathing with good stability, but it shows a rather low response time (Econsultancy 2020). A multiwalled carbon nanotubes (MWCNT) sensor was also proposed where apnea response and real tests were recorded in Balakrishnan et al. (2019). The study shows the efficiency of carboxylic functionalized MWCNTs deposit by inkjet printing for realization of sensitive and cost-effective humidity sensors. Vanadium oxide thin film on PDMS substrate have been reported to have good response time for inhalation and exhalation of 0.5 s (Liao et al. 2017).

In this contribution, we propose an easy processable and low-cost graphene oxide (GO) based sensors for breath monitoring. GO is layered chemically derived graphene containing oxygen functional groups (epoxy, hydroxyl, carboxyl and carbonyl). These functional groups make it an insulator with a resistance of hundreds of Mega Ohms (Al-Hamry et al. 2016). The hydrophilic nature of GO changes by the change of the oxygen quantity. Thus, the functional groups contained in GO influence the sensitivity of GO and its derivative, the so-called reduced graphene oxide (rGO). We propose to fabricate the sensor based on casting and optimal thermal annealing. By proper thermal annealing, the optimized surface properties and a suitable material sensitivity can be obtained. After optimization of preparation parameters, the film with the highest sensitivity to relative humidity is selected and tested for breath monitoring. In order to be able to store and analyze the measured data on long term, we realize an IoT solution able to visualize real time breath sensor data on a Mobile Application.

# 2 Experimental Investigations

## 2.1 Sensor Preparation

GO utilized in this work was purchased from Grphenea Inc. with the initial concentration of 4 mg/mL. The GO was diluted in deionized water to 1 mg/mL. Slight sonication is required to prevent agglomeration after dilution. The sensitive film was fabricated by dispersing GO on silver interdigitated electrodes. These were initially inkjet printed on the Kapton substrate and annealed at 170 °C. The line width and finger spacing of the interdigitated electrodes ar 200 μm. The total area of the sensor is 1 $cm^2$. The substrate was cleaned by isopropanol and dried by nitrogen. Then a drop of 10 μL of GO was casted onto the electrodes to form a continuous network of GO sheets. After deposition of GO, the GO was left overnight in a laboratory environment until it dried out. In order to partially reduce GO, the sensor was thermally annealed at two different temperatures i.e. 100 and 200 °C. After that, it was wired by support of silver paste to realize the connection to the measurement setup.

## 2.2 Experimental Methodology

Atomic force microscopy (AFM) was performed using a Keysight 5600LS system in order to investigate the morphology and structure of the deposited rGO. Humidity measurement was carried out by the Agilent 4292A impedance analyzer in the frequency range 40–110 MHz. The sensor was inserted in bottles filled with different saturated salt solutions having fixed-points of relative humidity of: 0, 11.3, 32.5, 52.3, 75.2, 84.3, and 97.3% RH, as shown in Fig. 1a. In order to monitor the breathing activity, an experimental setup was realized by using a flexible face mask to fix the rGO sensor as shown in Fig. 1b.

The high ohmic GO sensor was connected to an experimental measurement system using a voltage divider circuit to convert the sensor resistance variation to a voltage variation. The circuit, as shown in Fig. 1b, is supplied by a constant voltage battery source of 1.5 V, where GO is the sensor element. The oscilloscope (LeCroy waverunner 6050) is used to observe and remotely acquire data via a TCP/IP interface. An ad-hoc LabView virtual instrument has been specifically developed to plot and store the acquired data. A youth volunteer was asked to wear the mask and to breath for some minutes to register the breath activity.

Temperature control was carried out using a hot plate from room temperature until 100 °C and the resistance was registered by a Keithly 2636 sourcemeter.

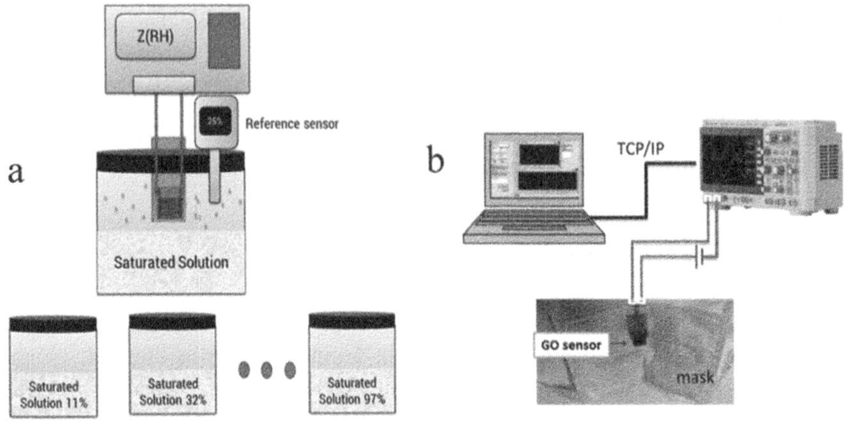


**Fig. 1** **a** rGO sensor and humidity measurements setup using saturated salt solutions and **b** breath monitoring measurement setup and schematic of the measurement setup

The feasibility of the IoT monitoring was realized using popular scripting language i.e. JavaScript and by standard tools for design of android mobile applications.

# 3 Results and Discussion

Graphene oxide is semi-insulating by nature having a resistance in the order of hundreds of MΩ. A reduction process is usually carried out e.g. by thermal annealing, photonic or by chemical reduction (Al-Hamry et al. 2016). In this work, we investigate the process of thermal annealing of GO to make so-called rGO. The range of reduction temperature was chosen so that the Kapton substrate withstands it in the range between 100 °C and 200 °C.

## *3.1 Physical Characterization*

In Fig. 2, UV-vis-NIR absorbance spectra of GO and rGO films are shown, which were annealed at different temperatures. The absorbance increases by increasing the annealing temperature. The spectra of GO show a peak around 260 nm, which indicates the $\pi - \pi^{\star}$ transition of $C - C$ bond. A shoulder at 320 nm is also observed that indicates the $n - \pi^{\star}$ plasmonic transition of $C = O$ bonds (Thakur and Karak 2012). The shoulder at 320 nm is still present in $rGO_{100\,^{\circ}C}$ absorption spectra and is totally disappeared in case of $rGO_{200\,^{\circ}C}$. This clearly indicates that the reduction process takes place and the elimination of oxygen containing groups start. In addition,

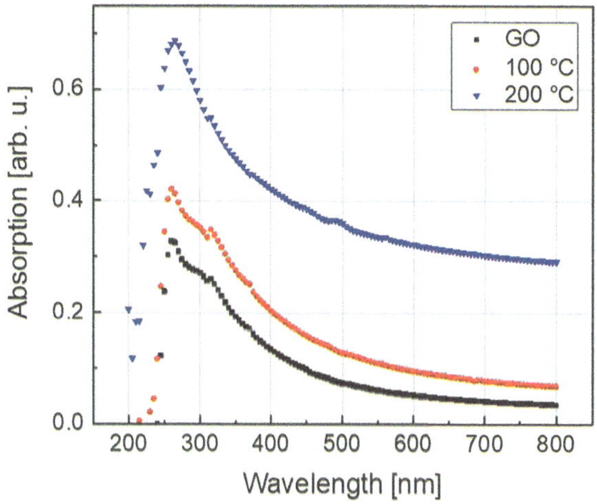


**Fig. 2** UV-vis-IR spectra of GO based films, untreated and annealed at 100° and 200 °C

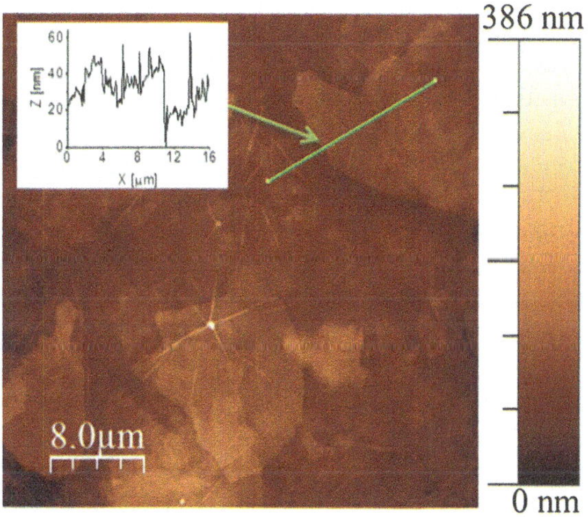


**Fig. 3** AFM images of the as deposited GO, Inset show the roughness in the marked line

for $rGO_{100\,°C}$, only partial reduction occurred where its resistance is still in order of hundreds of MΩ.

AFM images of GO are reported in Fig. 3. The images show the layered stacked structures of GO film where the roughness distribution within the film in the nanometer range. The stacked structure and porous-like topography enable water molecules to penetrate to interlayers of the film.

### 3.2 Electrical Measurements and Humidity Influence

Impedance Spectroscopy has shown an outstanding performance in the characterization of materials and measurement of humidity (Tetuyev and Kanoun 2006). The method allows measuring both real and imaginary part of conductivity at the same time and at different frequencies. This delivers a deep knowledge about the conductivity behavior.

In Fig. 4, the impedance versus relative humidity of GO, $rGO_{100\,°C}$ and $rGO_{200\,°C}$ is shown. The sensitivity behavior shows a dependency on reduction temperature. For GO sensor, the impedance decreases remarkably at 52% RH. Similarly, for rGO-100 °C, the impedance decreases more significantly at higher relative humidity values by magnitude of 500 times, relative to initial value at 0% RH. However, for the film reduced at higher temperature, $rGO_{200\,°C}$, the impedance shows a positive slope. It increases linearly with humidity increase. In this case, the surface becomes hydrophobic (Al-Hamry et al. 2016). Therefore, it is less sensitive to relative humidity.

Nyquist diagram $rGO_{100\,°C}$ is shown in Fig. 5. The behavior in the low humidity range (0–52%) is capacitive with high resistive parts. A slight decrease of the impedance due to the absorbed water molecules is observed. At these low humidity values, charge transfer between water molecules and GO/rGO layers occurs. However, it is not efficiently increasing the charge carriers of the film (Zhang et al. 2014a).

At high humidity (75–97%), the equivalent circuit consists of a capacitance representing the interaction of water and contacts and a resistance representing the intrinsic resistance of the film. For GO reduced at higher temperatures, by increase of water molecules the charge transfer is dominant and causes further increase of the impedance. For GO and low reduced films the formation of water thin films has the role of decreasing the impedance due to proton hopping (Zhang et al. 2014b). In addi-

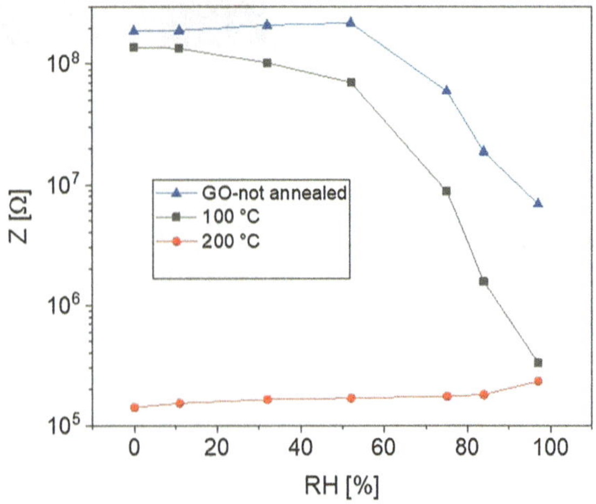


**Fig. 4** GO sensor dependency to relative humidity for non-annealed GO films and $rGO_{100\,°C}$ and $rGO_{200\,°C}$ films

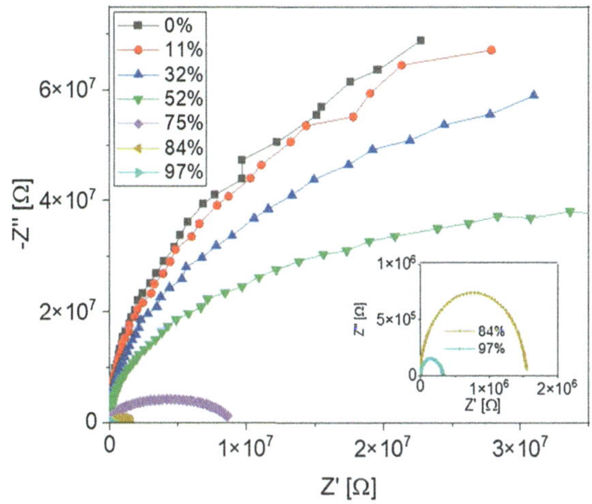


**Fig. 5** $GO_{100\,°C}$ dependency to relative humidity in Nyquist plot

tion, by increasing the reduction temperature, rGO films become more hydrophobic which prevents the water attraction to its surface and prevents making continuous water films. From the analysis of the humidity dependency tests, the $rGO_{100\,°C}$ was selected to realize a breath monitoring sensor. The advantage of $rGO_{100\,°C}$ is that it is insensitive to ambient room humidity and highly sensitive at high humidity. This makes it useful to detect breath. In addition, it is temperature independent as long as the temperature is lower than effective reduction temperature of GO. That means it remains semi-insulating and insensitive to room ambient temperature as well.

The temperature dependency of the GO film was tested in a range from 30 to 100 °C. Figure 6 shows that the resistance has values more than 10 MΩ up to 80 °C. Between 30 and 40 °C, the resistance value is more than 43 MΩ. This value is much higher than the resistance corresponding to 75% relative humidity, where the resistance starts to drop drastically by increasing the humidity.

Therefore, temperature dependency at temperatures below 40 °C can be neglected comparing to sensitivity of RH coming from breathing specially at room temperature range 20–40 °C.

## 3.3 *Breath Monitoring System and IoT Application*

### 3.3.1 Breath Measurement

The experimental results of breath monitoring are shown in Fig. 7. The amplitude varies according to the duration and the depth of the breath inhalation and exhalation. The acquired results show an amplitude variation in the range 0–1.22V that corresponds to a resistance variation of 5 MΩ approximately. The reported results in

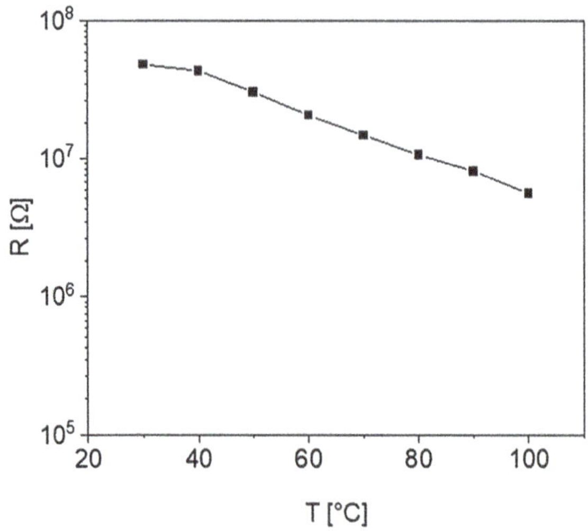


**Fig. 6** Temperature dependency of reduced graphene prepared by thermal annealing at 100 °C

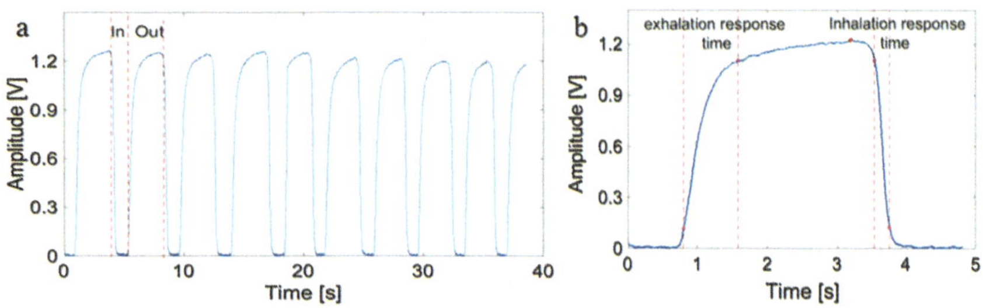


**Fig. 7** **a** Breathing measurement over several cycles and **b** one cycle showing the response and recovery time

Fig. 7a show the sensor signal for 40 s with a sampling rate 50 Hz at room temperature of approximately 23 °C.

In Fig. 7b, the response and recovery time by inhalation and exhalation are shown. This figure shows that the sensor response and recovery times are 248 ms and 784 ms, for inhalation and exhalation, respectively. These reaction times are lower than those reported in several investigations (Caccami et al. 2018; Liao et al. 2017), which indicates that the developed sensor is an ultrafast breath-monitoring sensor. This property owes to the hydrophilic nature of graphene oxide and the tunability of the humidity sensing properties. In Fig. 8, different breath patterns were registered at respiration rates which confirms the ability of the sensor to determine respiration rate and respiration depth. The graphs suggest that respiration rates from 12, 24 and 36 per minute for cases of deep, normal and fast breathing, respectively.

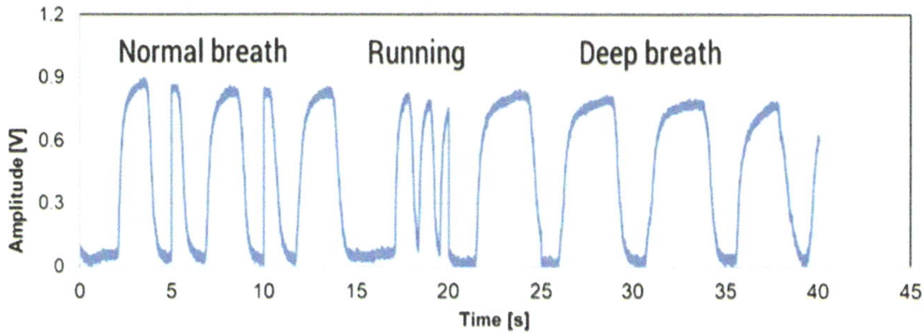


**Fig. 8** Detection of breath patterns

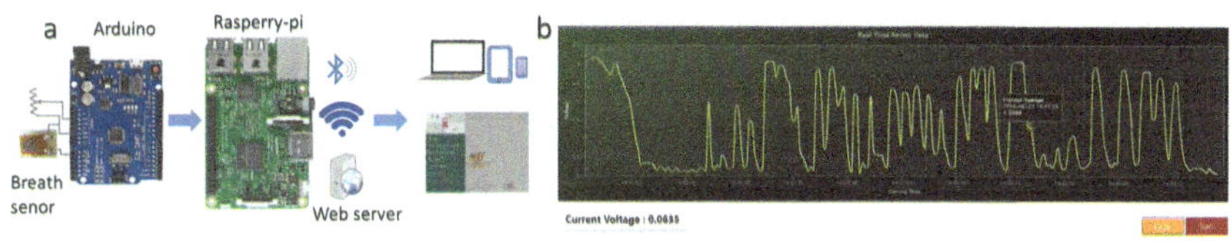


**Fig. 9** **a** Schematic of an IoT based mobile visualization and **b** Breathing tracking at different rates

### 3.3.2 IoT Application for Breath Monitoring

The GO based sensor was connected to Arduino in voltage divider with potentiometer of 5 MΩ to control the sensitivity, Fig. 9. Basically, the resistance of the sensor goes below 5 MΩ when the humidity is around 75%. At this point, according to voltage divider Eq. (1), the voltage at the input of the Arduino will be $\frac{1}{2}V_c$, where $V_c$ is 5 V from Arduino voltage port.

$$V_{in} = V_c \times \frac{R_{rGO100}}{R + R_{rGO100}} \tag{1}$$

The sensitivity to breath can be adjusted according to the ambient humidity by changing the potentiometer to avoid any background measurement.

Raspberry-pi device was used for communication via Bluetooth, Wi-Fi and internet. For data visualization from the breath sensor without delay, accuracy and speed are very important. For this purpose, a robotic framework (Johnny-Five) was used (Johnny-Five 2018) to get maximum speed for reading data from Arduino. The whole project was implemented with JavaScript and achieves maximum speed of 25 ms.

The sensor data can be visualized in different platforms i.e. web cloud and smart devices such as smart mobile phone and android tablets. Raspberry-pi, using Jonny-5 enabled by Javascript (Node.js), collects data from Arduino, process data then send data via Bluetooth and Wi-Fi to smart devices or web. Ionic3 hybrid mobile frame work was used for development of the mobile application. PubNub Data Stream Network (DSN) provides a global infrastructure and allows building and scale real-time apps and IoT devices to publish the sensor data to be read in the browser (PubNub 2018). Figure 9a shows the schematic of the data flow from sensor to data displaying using different paths. Figure 9b shows arbitrary respiration rhythms shown from a screen shot of an android device.

## 4 Conclusion

A preparation procedure was developed to synthetize a sensitive GO film based on reduced graphene oxide deposited on a flexible substrate. The GO film was reduced at a low temperature under 200 °C and shows an ultrahigh sensitivity of impedance to humidity, which is of the order of several magnitudes relative to initial value in the low humidity range. In the humidity range greater than 50%, the impedance is decreasing remarkably, so that it suitable for use in breath monitoring. The developed sensor was used to track the human breathing and shows good sensitivity and repeatability. The response time was found to be less than one second, i.e. 248 ms and 784 ms, for inhalation and exhalation, respectively. A mobile application for breath monitoring was developed for IoT based reading and visualization of data. As a sensor with very fast response time, it is of high potential for further use in real-time breath monitoring applications. It is flexible, repeatable, fast, delivers easy processable signals and is IoT compatible.

## References

Al-Hamry, A., Kang, H., Sowade, E., Dzhagan, V., Rodriguez, R. D., Müller, C., et al. (2016). Tuning the reduction and conductivity of solution-processed graphene oxide by intense pulsed light. *Carbon, 102*, 236–244.

Balakrishnan, V., Dinh, T., Foisal, A. -R. -M., Nguyen, T., Phan, H. -P., Dao, D. V., & Nguyen, N. -T. (2019). Paper-based electronics using graphite and silver nanoparticles for respiration monitoring. *IEEE Sensors Journal, 19*(24), 11784–11790.

Bouhamed, A., Rajendran, D., Frenzel, P., Zubkova, T., Al-Hamry, A., Miesel, D., Kamatchi, V., Ramalingame, R., Bautista-Quijano, J. R., Lang, H., Baumann, R. R., & Kanoun, O. (2021). Customizing hydrothermal properties of inkjet printed sensitive films by functionalization of carbon nanotubes. *Nanotechnology, 32*(10), 105708.

Caccami, M. C., Mulla, M. Y., Di Natale, C., & Marrocco, G. (2018). Graphene oxide-based radiofrequency identification wearable sensor for breath monitoring. *IET Microwaves, Antennas & Propagation, 12*(4), 467–471.

Econsultancy. (2020). 10 examples of the Internet of Things in healthcare. Retrieved June 22, 2020, from https://econsultancy.com/internet-of-things-healthcare/.

Johnny-Five. (2018). JavaScript robotics API documentation (Johnny-Five). Retrieved July 23, 2018, from http://johnny-five.io/api/.

Kanoun, O., Keutel, T., Viehweger, C., Zhao, X., Bradai, S., Naifar, S., Trigona, C., Kallel, B., Chaour, I., Bouattour, G., Khriji, S., El Houssaini, D., & Götz, M. (2018). Next generation wireless energy aware sensors for internet of things: A review. In *15th International Multi-Conference on Systems, Signals and Devices, SSD*, 19–22 March 2018. Tunisia: Hammamet.

Kim, I. D., Choi, S. J., Kim, S. J., & Jang, J. S. (2015). Exhaled breath sensors. In C. M. Kyung (Ed.), *Smart sensors for health and environment monitoring*. KAIST Research Series. Dordrecht: Springer.

Lau, H. -C., Yu, J. -B., Lee, H. -W., Huh, J. -S., & Lim J. -O. (2017). Investigation of exhaled breath samples from patients with alzheimer's disease using gas chromatography-mass spectrometry and an exhaled breath sensor system. *Sensors*, *17*(8), 1783.

Liao, F., Zhu, Z., Yan, Z., Yao, G., Huang, Z., Gao, M., et al. (2017). Ultrafast response flexible breath sensor based on vanadium dioxide. *Journal of Breath Research*, *11*(3), 36002.

Liu, H.-C., Chen, Y.-J., Lu, Y.-C., Wu, C.-L., Huang, W.-C., & Huang, J.-T. (2013). Monitoring Apnea in the elderly by an electromechanical system with a carbon nanotube-based sensor. *International Journal of Gerontology*, *7*(3), 147–151.

PubNub (2018). PubNub JavaScript V4 SDK 4.21.2. Retrieved July 21, 2018, from https://www.pubnub.com/docs/web-javascript/pubnub-javascript-sdk.

Righettoni, M., Tricoli, M., Gass, S., Schmid, A., Amann, A., & Pratsinis, S. E. (2012). Breath acetone monitoring by portable Si:WO3 gas sensors. *Analytica Chimica Acta*, *738*, 69–75.

Tetuyev, A., & Kanoun, O. (2006). Method of soil moisture measurement by impedance spectroscopy with soil type recognition for in-situ applications. *Technisches Messen*, *73*(7/8), 404–412

Thakur, S., & Karak, N. (2012). Green reduction of graphene oxide by aqueous phytoextracts. *Carbon*, *50*(14), 5331–5339.

Van Den Broek, J., Güntner, A. T., & Pratsinis, S. E. (2018). Highly selective and rapid breath isoprene sensing enabled by activated alumina filter. *ACS Sensors*, *3*(3), 677–683.

Zhang, D., Liu, A., Chang, H., & Xia, B. (2015). Room-temperature high-performance acetone gas sensor based on hydrothermal synthesized SnO2-reduced graphene oxide hybrid composite. *Journal RSC Advances*, *5*(4), 3016–3022.

Zhang, D., Tong, J., & Xia, B. (2014a). Humidity-sensing properties of chemically reduced graphene oxide/polymer nanocomposite film sensor based on layer-by-layer nano self-assembly. *Sensors and Actuators B: Chemical*, *197*, 66–72.

Zhang, D., Tong, J., Xia, B., & Xue, Q. (2014b). Ultrahigh performance humidity sensor based on layer-by-layer self-assembly of graphene oxide/polyelectrolyte nanocomposite film. *Sensors and Actuators B: Chemical*, *203*, 263–270.

# Impedimetric Detection of Human Interleukin 10 on Diazonium Salt Electroaddressed Gold Microelectrode Surfaces

**Michael Lee, Abdoullatif Baraket, Monique Sigaud, Ammar Al-Hamry, Nadia Zine, Olfa Kanoun, Joan Bausells, and Abdelhamid Errachid**

**Abstract** In this chapter, we describe the development and fabrication of gold microelectrodes based on silicon by silicon technology, for multiplex detection of cytokines. Cytokines have become a crucial biomarker for the identification of end-stage heart failure (ESHF) for patients during early phase of left ventricular assisted device (LVAD) implantation. The microelectrode device consists of three gold working microelectrodes that were activated and 4-aminophenylacetic acid (CMA) was electroaddressed onto individual gold WEs. The carboxylic acid functionalities of the diazotated aromatic amine were activated through carbodiimide chemistry and anti-interleukin-10 monoclonal antibodies (anti-IL-10 mAb) were immobilized onto the transducers surface. The interaction between the antibody-antigen (Ab-Ag) was characterized by electrochemical impedance spectroscopy (EIS). Here, Nyquist

M. Lee · A. Baraket · M. Sigaud · N. Zine · A. Errachid (✉)
Université de Lyon, Institut de Sciences Analytiques (ISA), Villeurbanne, France
e-mail: abdelhamid.errachid@univ-lyon1.fr

M. Lee
e-mail: michael.lee@isa-lyon.de

A. Baraket
e-mail: Abdoullatif.Baraket@inrae.fr

M. Sigaud
e-mail: monique.sigaud@isa-lyon1.fr

N. Zine
e-mail: nadia.zine@univ-lyon1.fr

J. Bausells
Instituto de Microelectronica de Barcelona, IMB-CNM (CSIC), Bellaterra, Barcelona, Spain
e-mail: joan.bausells@imb-cnm.csic.es

A. Al-Hamry · O. Kanoun
Chair for Measurement and Sensor Technology, Technische Universität Chemnitz, Chemnitz, Germany
e-mail: ammar.al-hamry@etit.tu-chemnitz.de

O. Kanoun
e-mail: olfa.kanoun@etit.tu-chemnitz.de

 
O. Kanoun and N. Derbel (eds.), *Advanced Sensors for Biomedical Applications*, Smart Sensors, Measurement and Instrumentation 38,
https://doi.org/10.1007/978-3-030-71225-9_7

plots have shown a stepwise variation due to the charge transfer resistance (Rct) between the Ab activated surfaces with the detection of the human IL-10. For early expression monitoring, commercial proteins of human IL-10 were analyzed between 1 pg/mL and 100 pg/mL. The protein concentrations within the linear range of 1–50 pg/mL were detected and these values formulated a sensitivity of 0.008 $(pg/mL)^{-1}$ ($R^2 = 0.9840$). These preliminary results demonstrated that the developed biosensor was sensitive to the detection of human IL-10 and the calculated limit of detection (LOD) was measured at 0.156 $(pg/mL)^{-1}$. To validate the biosensors response, the experiment was repeated several times on different gold working WEs by applying the same conditions. The overall relative standard deviation percentage (% R.S.D.) was 4.9% which demonstrates the successful fabrication for the detection of human IL-10 through diazonium salt electroreduction.

**Keywords** Biosensor · Diazonium salt · Interleukin-10 · Electrochemical impedance spectroscopy · Cyclic voltammetry

## 1 Introduction

In recent years, there has been a phenomenal growth in the field of biosensor development. The requirement of chemically-modified surfaces is of critical importance for the immobilization of the biological material on the transducers surface. Such development is essential in order to attain high sensitivity and selectivity for detection during biorecognition processes (Mrksich and Whitesides 1995; Turner 2000). The formation of functional groups (amino, carboxyl, etc.) in silicon chemistry applying insulating materials such as silicon dioxide ($SiO_2$) (Bras et al. 2004; Zhang et al. 2010), silicon nitride ($Si_3N_4$) (Caballero et al. 2012; Stine et al. 2007), aluminum oxide ($Al_2O_3$) (Lee et al. 2014), and hafnium oxide ($HfO_2$) (Lee et al. 2012) requires silane chemistry. For conducting materials such as gold, one of the most commonly used techniques is based on alkanethiol monolayers (Baraket et al. 2014; Zamfir et al. 2011). This technique has been extensively applied through self-assembled monolayers (SAMs) (Love et al. 2011) followed by carbodiimide chemistry. The SAMs can be spatially controlled using microcontact printing ($\mu$CP) (Xia and Whitesides 1998) with structures observed at the nanometric scale (Renault et al. 2003).

Also, the electrochemical formation of polymers (for e.g. polypyrrole and polyaniline) has been applied for sensing applications (Biloivan et al. 2006; Cosnier 2003).

The modification of surfaces through diazonium salts was first introduced by Delamar et al. (1992). In this method, the diazonium salt is electrochemically reduced by cyclic voltammetry (CV) and this creates an aryl centered radical due to the spontaneous elimination of di-nitrogen. The resulting aryl radical can then form a covalent bond with conducting and semi-conducting surfaces (Polsky et al. 2008b). The modification of electrodes through diazonium salts is a promising alternative to conventional techniques as previously described (e.g. silanes and thiols). This has been demonstrated on a wide range of surfaces such as gold (Bernard et al. 2003;

Corgier et al. 2009), glassy carbon (Liu et al. 2011; Polsky et al. 2007), semiconductors (Pinson and Podvorica 2005; Stewart et al. 2004), screen printed electrodes (Yáñez-Sedeño et al. 2018), carbon nanotubes (CNTs) (Bahr et al. 2001; Lee et al. 2004), CVD graphene (Eissa et al. 2015), graphene and reduced graphene oxide composites (Ott et al. 2018).

Due to its simplicity, rapidity and versatility, diazonium salt grafting and functionalization is widely used to attach organic layer on conductive substrate, covalently immobilize biomolecules and nanomaterials or self-assembling of nanoparticle (Yáñez-Sedeño et al. 2018). Aryldiazonium chemistry have been utilized for functionalization of biosensor (Bagheryan et al. 2016) as well as cell immobilization (Polsky et al. 2008a). Aryldiazonium covalent chemistry on different surfaces e.g. graphene was achieved by self-assembly of thin layer of n-alkaline making masking monolayer and by EC potential aryldiazonium is diffused through them to generate aryl radicals on the surface (Tahara et al. 2018). Grafting end-moieties -OH, -COOH, and -NH2 commercial surfactant to the surface of electrodes via aryldiazonium slats was also demonstrated to form protective layer i.e. gold and metallic chromium thin film (Tahara et al. 2018). Self-assembly of Fréchet-type dendron nanoparticles (FDNs) nanoparticles (Krishnakumar and Gopidas 2017) and gold nanostructure (Betelu et al. 2016) was carried out by covalent functionalization using different aryldiazonium derivatives.

Corgier et al. (2005) conjugated 4-carboxymethylaniline to IgG antibodies and diazotated the diazonium salt-modified antibody onto screen-printed graphite electrode arrays (Corgier et al. 2005). Applying the same strategy, Polsky et al., immobilized different antibodies onto glassy carbon electrodes (GCE) and gold electrode arrays for the electrochemical detection of cytokines (Polsky et al. 2008b). However, a concern of this technique is due to the reaction conditions applied during diazotation. The conversion requires HCl (20 mM) and $NaNO_2$ (20 mM) which may not be compatible with all proteins (Radi et al. 2009). Therefore, a different approach has been applied that first diazotates the diazonium for covalent grafting onto the substrate surface. This is then followed by linker groups (for e.g., glutaraldehyde and carbodiimides) for the immobilization of, for example, DNA (Harper et al. 2007) and proteins (Bello-Gil et al. 2014). Despite different drawbacks such as creation of multilayers, side reactions and surface heterogeneities for some surfaces, potential solution have been developed (Jiang et al. 2017; Yáñez-Sedeño et al. 2018), such as mixed monolayers, control over spatial distribution and the structure of mixed layer.

Left-ventricular assisted devices (LVAD) implanted into patients suffering from end-stage heart failure (ESHF) has been established as an effective bridge to heart transplantation (HT) by enabling the recovery of adequate cardiovascular hemodynamic function (Caruso et al. 2010). During the first one month, patients are highly susceptible to multiple organ failure syndrome (MOFS) and as a consequence, ultimately lead to deaths. MOFS is influenced by the degree of the immunoinflammatory response, independent of the presence of infection (Caruso et al. 2010). The inflammatory response is analyzed by cytokine mediators and early release of pro- and anti-inflammatory cytokines have been observed in patients with a high risk of MOFS (Beger and Rau 2007). Standard techniques such as enzyme-linked immunosorbent assays (ELISA) are implemented to the quantification of cytokines, though immobilization, bio-conjugation, wascytokin h steps, and quantification requires increased time and input from qualified personnel. Point of care devices utilizing biomarker or HF management have been developed (Tripoliti et al. 2020). In regards of biosensors, different approaches were investigated to detect IL-10. Impedimetric diazonium modified IL-10 and anti.IL1b were electrodeposited on gold substrate (Baraket et al. 2017). Moreover, Label-free capacitance impedimetric immunosensor based on the immobilization of the human monoclonal anti-IL-10 mAb onto polypyrrole (PPy)-modified silicon nitride ($Si_3N_4$) substrates) was developed (Nessark et al. 2020).

For chronic heart failure (CHF) patients that have underwent LVAD implantation, Caruso et al. (2010) have assessed IL-10 with other varying cytokines for patients at high risk for MOFS during the first month post-LVAD implantation. Human IL-10 plasma levels for 23 LVAD implanted patients were between the ranges of 0–1.558 pg/mL in the first 30 days of LVAD support using ELISA. The authors analyzed samples within certain time frames to quantify exactly when patients were susceptible to higher levels of inflammation after surgery. Before LVAD implantation, patients obtained minimal IL-10 inflammation (pre-implant) due to the circulating cytokine (1-month survivors, 1.8 pg/mL and non-survivors, 5.6 pg/mL). Following LVAD implantation, plasma samples were taken after 4 h. At this critical stage, early expression of human IL-10 peaked in comparison with other plasma samples analyzed within the 30 day period for survivors and non-survivors. Levels for IL-10 non-survivors (177.8 pg/mL) were also higher than that of the survivors (56.2 pg/mL). Therefore, the authors have established that the elevated IL-10 levels in parallel with other cytokines can play an important role in the development of adverse events early after LVAD implantation.

In this chapter, we demonstrate the electrochemical behavior of gold microelectrodes through bare, activated, and 4-aminophenylacetic acid (CMA) functionalized surfaces using CV and electrochemical impedance spectroscopy (EIS) measurements. The results demonstrate the successful functionalization of the gold microelectrode through electrochemical methods. Afterwards, a biosensor was fabricated with the immobilization of anti-IL-10 monoclonal antibodies (mAbs) onto gold microelectrodes through 4-aminophenylacetic acid (CMA) electroaddressing. Here, the detection of human IL-10 and its interferences were analyzed.

## 2 Experimental Investigations

### *2.1 Materials and Techniques*

Chemicals, materials, and reagents: All chemicals were purchased from Sigma-Aldrich, France apart from 4-aminophenylacetic acid (CMA) (98%) which was purchased from Acros Organics, France. PDMS (Sylgard 184) was purchased from Dow Corning, France. The anti-human IL-10 mAb, recombinant human (rh) IL-10, rh IL-

8, and rh tumor necrosis factor-$\alpha$ (TNF-$\alpha$) were purchased from R&D Systems, France.

**Imaging**: Optical images were taken by optical microscopy (Leica EZ4D, Spain).

## 2.2 *Fabrication of Gold Microelectrodes*

The fabrication of the gold microelectrodes was developed at the CNM, Barcelona. Here, p type silicon < 100 > with a diameter of 100 mm and a nominal thickness of 525 μm was applied. The process began with thermal oxidation of silicon to grow a thick oxide layer (8000 Å). A photoresist resin layer was deposited in order to pattern the surface. To improve the adherence of gold microelectrodes a layer of Titanium (Ti) (1 nm) was applied as an intermediate layer that was followed with a sputtered layer of gold (250 nm). Afterwards, a photoresist resin was spin coated onto the surface and exposed to UV through the developed microelectrodes photomask. This was proceeded with plasma-enhanced chemical vapor deposition (PECVD) with two layers consisting of $SiO_2$ (4000 Å) and $Si_3N_4$ (4000 Å), respectively. These layers acted as passivation layers. A second photolithographic step was made to open passivation on the gold working microelectrodes (300 μm × 300 μm) and the corresponding electrical pads. After this step, the fabrication of the gold microelectrodes on silicon was complete. The microelectrodes were then bonded to a printed circuit boards

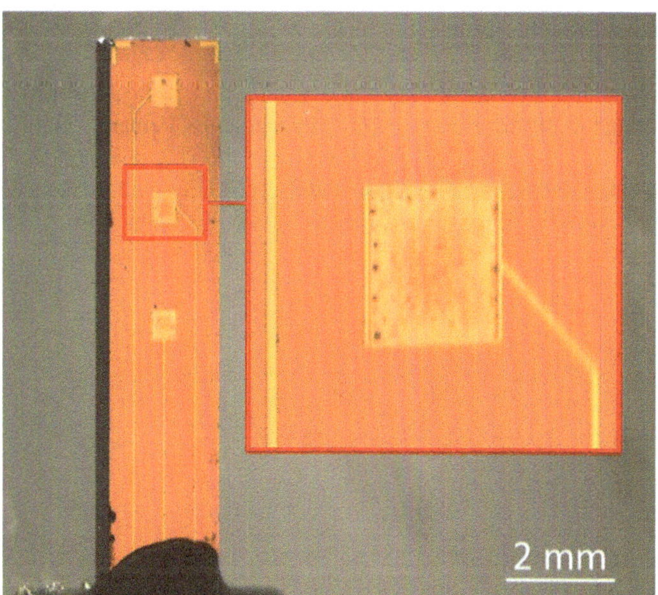


**Fig. 1** Optical image of the gold microelectrodes developed on silicon by silicon technology

(PCB) and the electrical connections were made by wire bonding (K&S 4500 Series Wire Bonders) in the usual manner. Finally, the microelectrodes were encapsulated by passivation with an epoxy resin (EPO-TEK H 70 E-2 LC, France). The developed gold microelectrodes are shown in Fig. 1.

## 2.3 Preparation of the Gold Microelectrode

The gold microelectrodes based on silicon were cleaned by rinsing with acetone and then thoroughly rinsed with DI water. They were then dried under a stream of nitrogen. The surface was then cleaned of organic contaminants using a UV/Ozone cleaner (UV/O3 Procleaner$_{TM}$, NanoAndMore GmbH, Germany) for 30 min.

## 2.4 Diazonium Electroaddressing

The gold microelectrode was rinsed in ethanol and immersed in a glass electrochemical cell for the electroaddressing using CMA. Here, the device was connected to the potentiostat (VMP3, EC-Lab, France) with an external reference electrode (Radiometer (REF 421)) and counter electrode (Radiometer platinum wire (M231PT)) was applied.

For the diazotation of 4-aminophenylacetic acid (CMA), 1 mL of a 10 mM solution of the aromatic amine was made-up in 1 M aqueous HCl/ethanol 50:50 (v/v). This was then mixed with 50 μL of an ice-cold solution of sodium nitrite (10 mg/mL) in water/ethanol 50:50 (v/v) for 5 min at 0 °C under magnetic stirring. The solution was then added to 20 mL of ice-cold 0.1 M aqueous HCl/ethanol 50:50 (v/v). This produced a 0.48 mM solution of the diazonium salt. This is a modification of the electrode preparation reported by Beger and Rau (2007), Squillace et al. (2017). The final solution was kept at 0 °C and immediately used. The reductive adsorption of the

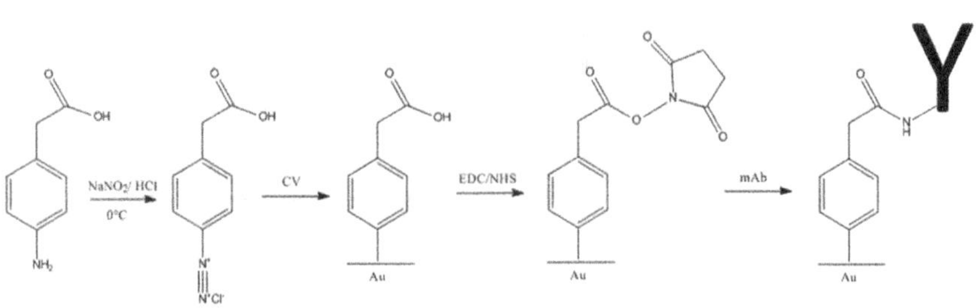


**Fig. 2** Reaction scheme for the electroaddressing of CMA onto a gold substrate, followed by activation with EDC/NHS, and the immobilization of the anti-IL-10 mAb

salt onto the electrode was achieved by four scans of CV between 0.3 and −1.1 V at 80 mV/s. The development of the biosensor can be observed in Fig. 2.

## 2.5 *Biorecognition of Human IL-10*

The gold microelectrode was rinsed in phosphate buffered saline (PBS) and a solution of EDC (0.4 M) and NHS (0.1 M) was made up in PBS (pH 7.4 at 10 mM) and incubated in the cell for 1 h at room temperature. After this time, the cell was rinsed with PBS.

The anti-human IL-10 was diluted to a concentration of 10 μg/mL in PBS. The modified microelectrode was incubated in the solution for 1 h at 4 °C. The substrate was then rinsed with PBS. The surface was then blocked with a 0.1% solution of ethanolamine in PBS for 20 min at room temperature. Afterwards, the microelectrodes were rinsed in PBS.

The human IL-10 was diluted in PBS to concentrations of 1, 5, 10, 15, 20, 30, 50, and 100 pg/mL, respectively. The modified gold-anti-IL-10 mAb microelectrode was first measured by EIS followed by increasing concentrations of human IL-10. Here, and after each measurement, the human IL-10 solution was incubated on top of the gold microelectrodes for 30 min at 4 °C. The substrate was then rinsed with PBS.

CV measurements were made at: potential of 0.6 V to −0.2 V, scan rate at 100 mV/s, and 3 cycles. EIS measurements were made at: potential of 0.228 V, frequency at 200 kHz to 200 mHz, and a sinus amplitude of 25 mV. The electrolyte was 5 mM of K3(Fe(CN)6/K4(Fe(CN)6 prepared in 400 mL of PBS solution.

# 3 Results and Discussion

## 3.1 *Electrochemical Measurements*

### 3.1.1 Biosensor Development

After packaging of the biosensor, the device was connected to a potentiostat and its electrochemical properties were observed. In Fig. 3a, the bare gold microelectrode shows characteristic CV peak currents for the redox reaction with ferri/ferrocyanide solution (Fig. 3a (black)). After UV/O3 cleaning, the peak currents increased which demonstrates an improved quality of the active surface (Fig. 3a (red)). After diazotation of CMA onto the gold microelectrode surface, the peak currents decreased which demonstrates the functionalization of the gold surface with CMA. This is due to the blocking behavior of the CMA on the modified gold microelectrode (Fig. 3a (blue)) (Baraket et al. 2013).

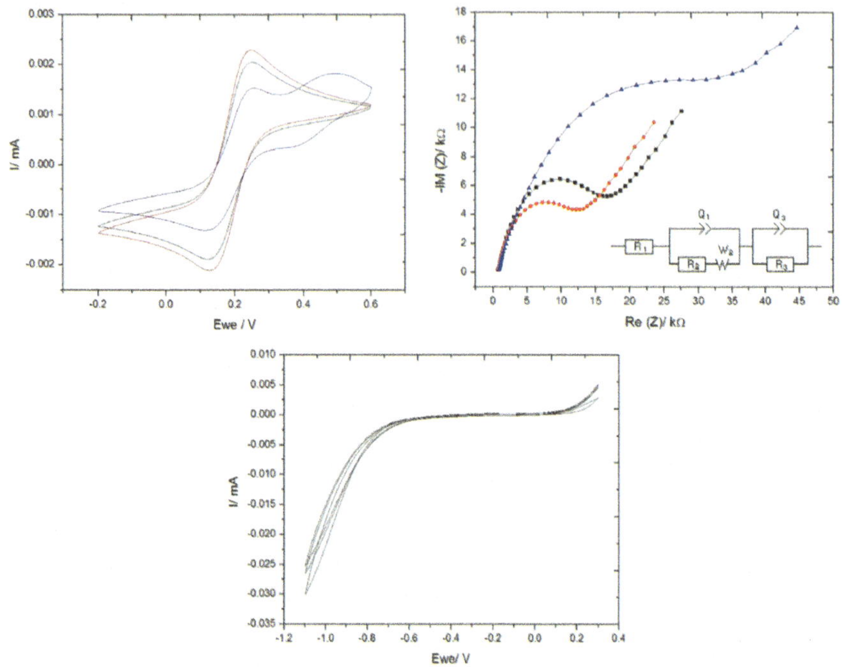


**Fig. 3** **a** CV scans and **b** EIS measurements of: (black) Bare Au, (red) after UV/O3 activation, and (blue) after CMA electroaddressing. **c** CV scan for the electroaddressing of CMA onto activated gold microelectrodes with three scans of CV between 0.3 and −1.1 V at 80 mV/s

For EIS measurements (Fig. 3b), the results were comparable to the CV measurements. The charge transfer resistance (Rct) of bare gold was measured at 13.6 kΩ (Fig. 3b "■") and a decrease in the $R_{ct}$ was observed after UV/$O_3$ activation at 11.0 kΩ (Fig. 3b "●"). This demonstrates that a cleaner surface has been obtained. After CMA electroaddressing, the $R_{ct}$ increased to 37.1 kΩ which corroborates with CV that the gold microelectrode surface was functionalized with CMA (Fig. 3b ("▲")). The values of the fitting parameters can be found in Table 1. The equivalent electrical circuit ($R_1 + Q_1/(R_2 + W_2) + Q_3/R_3$) was applied for the simulation and it formulated the best fit for the data. Here, the components can be explained as follows: $R_1$ expresses the electrolyte solution ($R_s$). $Q_1$ is the constant phase element (CPE) ($Q_{\sigma_1}$ and $Q_{\alpha_1}$) that is parallel with $R_2$ for the first semi-circle which is in series with $W_2$ ($S_2$) which is the Warburg resistance, $Q_3$ is the CPE ($Q\sigma_3$ and $Q\alpha_3$) that is parallel with $R_3$ for the second semi-circle and is designated as the $R_{ct}$. For the Zfit, the Nyquist plots were observed with Randomize + Simplex method, with randomize stopped on 10,000 iterations and the fit stopped on 5,000 iterations.

In Fig. 3c, the electroaddressing of CMA onto the gold WEs is shown. Here, a typical electroreduction of the CMA leads to the elimination of the nitrogen molecule and it produces an aryl radical (Pinson and Podvorica 2005; Radi et al. 2009). This

**Table 1** EIS fitting parameters the characterization of the gold microelectrodes development

| | $R_1$ (kΩ) | $Q_1$ ($\mu$F$s^{a-1}$) | $R_2$ (kΩ) | $s_2$ (kΩ$s^{-0.5}$) | $Q_3$ (nF$s^{a-1}$) | $R_3$ (kΩ) | $X^2$ |
|---|---|---|---|---|---|---|---|
| (1) Bare Au "■" | $0.3 \pm 5.9 \times 10^{-3}$ | $0.4 \pm 0.2$ | $4.1 \pm 0.6 \times 10^{-3}$ | $47.5 \pm 695.2 \times 10^{-3}$ | $128.7 \pm 1.9$ | $13.6 \pm 64.9 \times 10^{-3}$ | 0.002 |
| (2) $UV/O_3$ of Au "●" | $0.7 \pm 2.9 \times 10^{-3}$ | $3.1 \pm 0.7$ | $7.9 \pm 1.1 \times 10^{-3}$ | $39.6 \pm 2.2 \times 10^{-3}$ | $169.1 \pm 1.4$ | $11.0 \pm 267.5 \times 10^{-3}$ | 0.006 |
| (3) After CMA "▲" | $10.2 \pm 3 \times 10^{-3}$ | $35.3 \pm 0.3$ | $7.5 \pm 1.4 \times 10^{-3}$ | $16.4 \pm 3.3 \times 10^{-3}$ | $734.5 \pm 8.5$ | $37.1 \pm 222.6 \times 10^{-3}$ | 0.002 |

radical then forms a covalent bond between the aryl group and the electrode surface (Betelu et al. 2016; Mesnage et al. 2012; Pinson and Podvorica 2005).

A decrease during the second and third voltammetric cycles has demonstrated that the surface has begun to be saturated with the CMA molecules. This is evidenced with the calculation of the percentage coverage area of the CMA that was analyzed by applying the relation (Corgier et al. 2005; Tlili et al. 2005):

$$\theta = 100 \times \left(1 - \frac{R_{\text{before silanization}}}{R_{\text{after silanization}}}\right) \tag{1}$$

where, the $R$ before and after silanization were obtained with the fitting program, respectively. This was calculated at 70.4%. This percentage demonstrates that electroaddressing of CMA was successful on the activated gold WEs using the applied conditions.

### 3.1.2 Detection of Human IL-10

Figure 4a, shows the Nyquist plots for the detection of human IL-10 at increased concentrations (1–100 pg/mL). Here, the Nyquist plot showed a stepwise variation with increasing Rct which demonstrates the successfully biorecognition of the Ab-Ag complex on the developed gold WEs through CMA functionalization.

Afterwards, the data was normalized using Z-fit and the equivalent electrical circuit $(R_1 + Q_1/(R_2 + W))$ was applied for the simulation and it formulated the best fit for the data (Table 2).

For biological measurements, for e.g., the detection of antigens, the given $R_1$ values of the Ab followed by increasing concentrations of the Ag can be applied to calculate the linearity, sensitivity, etc. for the applied biosensor. This is accomplished

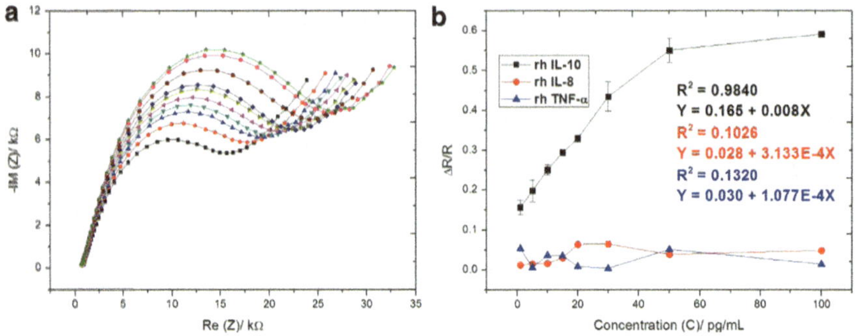


**Fig. 4** **a** Nyquist plot for the detection of human IL-10 and **b** normalized data for the detection of human IL-10 and interferences study using human IL-8 and human TNF-$\alpha$

**Table 2** Fitting parameters from the applied equivalent circuit for the detection of human IL-10

| | $R_1$ (kΩ) | $Q_2$ ($\mu$F$s^{a-1}$) | $R_2$ (kΩ) | $S_2$ (kΩ$s^{-0.5}$) | $X^2$ |
|---|---|---|---|---|---|
| Anti-IL-10 mAb | $710.2 \pm 0.5$ | $0.8 \pm 0.6 \times 10^{-3}$ | $21.5 \pm 25.1 \times 10^{-3}$ | $3.6 \pm 103.2 \times 10^{-3}$ | 0.027 (10 μg/mL) |
| rh IL-10 | $683.7 \pm 0.4$ | $0.7 \pm 0.4 \times 10^{-3}$ | $24.9 \pm 21.1 \times 10^{-3}$ | $4.0 \pm 79.6 \times 10^{-3}$ | 0.032 (1 pg/mL) |
| rh IL-10 | $682.4 \pm 0.3$ | $0.6 \pm 0.3 \times 10^{-3}$ | $25.8 \pm 21.0 \times 10^{-3}$ | $6.0 \pm 78.1 \times 10^{-3}$ | 0.033 (5 pg/mL) |
| rh IL-10 | $660.3 \pm 0.4$ | $0.6 \pm 0.2 \times 10^{-3}$ | $26.9 \pm 21.7 \times 10^{-3}$ | $9.1 \pm 76.8 \times 10^{-3}$ | 0.039 (10 pg/mL) |
| rh IL-10 | $732.0 \pm 0.4$ | $0.6 \pm 0.1 \times 10^{-3}$ | $27.9 \pm 21.4 \times 10^{-3}$ | $10.7 \pm 74.7 \times 10^{-3}$ | 0.037 (15 pg/mL) |
| rh IL-10 | $633.1 \pm 0.4$ | $0.5 \pm 0.1 \times 10^{-3}$ | $28.6 \pm 21.4 \times 10^{-3}$ | $13.5 \pm 73.3 \times 10^{-3}$ | 0.040 (20 pg/mL) |
| rh IL-10 | $650.3 \pm 0.4$ | $0.5 \pm 0.1 \times 10^{-3}$ | $30.9 \pm 23.0 \times 10^{-3}$ | $16.7 \pm 72.4 \times 10^{-3}$ | 0.041 (30 pg/mL) |
| rh IL-10 | $641.6 \pm 0.4$ | $0.5 \pm 0.2 \times 10^{-3}$ | $33.4 \pm 24.0 \times 10^{-3}$ | $22.0 \pm 69.8 \times 10^{-3}$ | 0.044 (50 pg/mL) |
| rh IL-10 | $620.7 \pm 0.4$ | $0.5 \pm 0.2 \times 10^{-3}$ | $34.3 \pm 24.3 \times 10^{-3}$ | $24.5 \pm 68.6 \times 10^{-3}$ | 0.045 (100 pg/mL) |

using the equation:

$$\frac{R_{Ag} - R_{Ab}}{R_{Ab}} \tag{2}$$

In Fig. 4b, the $R_1$ and the concentration ($C$) of human IL-10 produced a linear relationship towards its specific antibody. The linear dynamic range was observed between 1 pg/mL to 50 pg/mL. At 100 pg/mL, the biosensor was saturated. Here, a sensitivity was observed at 0.008 (pg/mL)$^{-1}$ with $R_2 = 0.9840$.

Applying the same conditions for the immobilization of anti-IL-10 mAb, the biosensor was analyzed for interferences using two inactive cytokines: human IL-8 and TNF-$\alpha$, respectively. These are other prevalent biomarkers that are relevant to LVAD patients. Here, the same concentrations (1–100 pg/mL) were applied. Reduced $\Delta R/R$ values that were significant to human IL-8 (red) at a sensitivity of $3.133\times10^{-4}$ $(\text{pg/mL})^{-1}$ ($R_2 = 0.1026$) and human TNF-$\alpha$ (blue) at a sensitivity of $1.077\times10^{-4}$ $(\text{pg/mL})^{-1}$ ($R_2 = 0.1320$) demonstrates that the sensitivity was much lower when compared to human IL-10. This demonstrates that the developed biosensor was sensitive and specific to only human IL-10.

The theoretical limit of detection (LOD) based on the standard deviation (SD) of the response and the slope (S) was calculated using the formula LOD = 3(SD/S) (Białk-Bielińska et al. 2011; Radi et al. 2009). The LOD was calculated at 0.156 $(\text{pg/mL})^{-1}$.

To validate the biosensors response, the experiment was repeated several times on different gold working WEs by applying the same conditions. The overall relative standard deviation percentage (%R.S.D.) for the detection of human IL-10 was 4.9%. This proves the reproducibility of the biosensor.

## 4 Conclusions

In this study, we realized a surface functionalization of gold microelectrodes by formulating a monolayer of CMA by electroaddressing. Surface modification on gold working microelectrodes has enabled covalent bonding with the mAb and demonstrated the effectiveness through electrochemical measurements. EIS has shown that a detection can be realized with varying human IL-10 concentrations in the picogram range. Here, a linear dynamic range was observed between 1 to 50 pg/mL, which is suitable for early phase detection. The level of interference attributable to non-specific binding of inactive proteins showed that the developed biosensor was selective and sensitive to only human IL-10.

Work is on-going to optimize all conditions, while future applications within biological samples (for e.g., plasma) will determine the LOD, linearity, response time, and sensitivity using these gold working microelectrodes. The increment of the dynamic range to larger concentrations can express the high risk of MOFS as a direct cause to implantation. Therefore, nanomaterials will be applied to improve the dynamic range.

**Acknowledgements** We acknowledge the funding through the European Union's Horizon 2020 and innovation programme entitled (An integrated POC solution for non-invasive diagnosis and therapy monitoring of Heart Failure patients, KardiaTool) under grant agreement No. 768686. This work partially supported by POC4Allergies # ERA-PerMed-2019 and PHC PROCOPE #40544QH projects.

## References

Allongue, P. M., Delamar, M., Desbat, B., Fagebaume, O., Hitmi, R., Pinson, J., et al. (1997). *Journal of the American Chemical Society*, *119*, 201–207.

Bagheryan, Z., Raoof, J.-B., Golabi, M., Turner, A. P. F., & Beni, V. (2016). *Biosensors and Bioelectronics*, *80*, 566–573.

Bahr, J. L., Yang, J., Kosynkin, D. V., Bronikowski, M. J., Smalley, R. E., & Tour, J. M. (2001). *Journal of the American Chemical Society*, *123*, 6536.

Baraket, A., Lee, M., Zine, N., Sigaud, M., Yaakoubi, N., Trivella, M. G., et al. (2013). *Sensors and Actuators, B*, *189*, 165–172.

Baraket, A., Lee, M., Zine, N., Yaakoubi, N., Trivella, M. G., Elaissari, A., et al. (2014). *Sensors and Transducers SI*, *27*, 15–21.

Baraket, A., Lee, M., Zine, N., Caruso, R., Trivella, M.G., & Errachid, A. (2017). *International Journal of Cardiovascular Research*, *06*.

Beger, H., & Rau, B. M. (2007). Severe acute pancreatitis: Clinical course and management. *World Journal of Gastroenterology*, *13*, 5043–5051.

Bello-Gil, D., Maestro, B., Fonseca, J., Feliu, J. M., Climent, V., & Sanz, J. M. (2014). *PLOS One*, *9*,

Bernard, M.-C., Chaussé, A., Cabet-Deliry, E., Chehimi, M. M., Pinson, J., Podvorica, F., et al. (2003). *Chemistry of Materials*, *15*, 3450–3562.

Betelu, S., Tijunelyte, I., Boubekeur-Lecaque, L., Ignatiadis, I., Ibrahim, J., Gaboreau, S., et al. (2016). *The Journal of Physical Chemistry C*, *120*, 18158–18166.

Białk-Bielińska, A., Siedlewicz, G., Stepnowski, P., Pazdro, K., Fabiańska, A., & Kumirskaa, J. (2011). *Analytical Methods*, *3*, 1371–1378.

Biloivan, O. A., Verevka, S. V., Dzyadevych, S. V., Jaffrezic-Renault, N., Zine, N., Bausells, J., et al. (2006). *Materials Science and Engineering C*, *26*, 574–577.

Bourigua, S., Hnaien, M., Bessueille, F., Lagarde, F., Dzyadevych, S., Maaref, A., et al. (2014). *Biosensors and Bioelectronics*, *26*, 1278.

Bras, M., Dugas, V., Bessueille, F., Cloarec, J. P., Martin, J. R., Cabrera, M., et al. (2004). *Biosensors and Bioelectronics*, *20*, 797–806.

Caballero, D., Samitier, J., Bausells, J., & Errachid, A. (2009). *Small*, *5*, 1531–1534.

Caballero, D., Martinez, E., Bausells, J., Errachid, A., & Samitier, J. (2012). *Journal Analytica Chimica Acta*, *720*, 43–48.

Caballero, D., Fumagalli, L., Teixidor, F., Samitier, J., Errachid, A. (2013). *Sensors and Actuators, B*, *177*, 1003–1009.

Caruso, R., Trunfio, S., Milazzo, F., Campolo, J., De Maria, R., Colombo, T., et al. (2010). *ASAIO Journal*, *56*, 313–318.

Cosnier, S. (2003). *Electroanalysis*, *17*, 1701–1715.

Corgier, B. P., Marquette, C. A., & Blum, L. J. (2005). *Journal of the American Chemical Society*, *127*, 18328.

Corgier, B. P., Bellon, S., Anger-Leroy, M., Blum, L. J., & Marquette, C. A. (2009). *Langmuir*, *25*, 9619–9623.

Delamar, M., Hitmi, R., Pinson, J., & Savéant, J.-M. (1992). *Journal of the American Chemical Society*, *114*, 5883–5884.

Eissa, S., Jimenez, G. C., Mahvash, F., Guermoune, A., Tlili, C., Szkopek, T., et al. (2015). *Nano Research*, *8*, 1698–1709.

Garcia-Cruz, A., Zine, N., Sigaud, M., Lee, M., Marote, P., Lanteri, P., et al. (2014). *Microelectronic Engineering*, *121*, 167–174.

Geetha, S., Raoa, C. R. K., Vijayana, M., & Trivedi, D. C. (2006). *Analytica Chimica Acta*, *568*, 119–125.

Harper, J. C., Polsky, R., Wheeler, D. R., Dirk, S. M., & Brozik, S. M. (2007). *Langmuir*, *23*, 8285–8287.

Harper, J. C., Polsky, R., Wheeler, D. R., & Brozik, S. M. (2008). *Langmuir*, *24*, 2206–2211.

Jiang, C., Moraes Silva, S., Fan, S., Wu, Y., Alam, M. T., Liu, G., et al. (2017). *Journal of Electroanalytical Chemistry, 785*, 265–278.
Krishnakumar, S., & Gopidas, K. R. (2017). *Langmuir the ACS Journal of Surfaces and Colloids, 33*, 1162–1170.
Lee, C.-S., Baker, S. E., Marcus, M. S., Yang, W., Eriksson, M. A., & Hamers, R. J. (2004). *Nano Letters, 4*, 1713–1716.
Lee, M., Zine, N., Baraket, A., Zabala, M., Campabadal, F., Caruso, R., et al. (2012). *Sensors and Actuators, B, 175*, 201–207.
Lee, M., Baraket, A., Zine, N., Zabala, M., Campabadal, F., Caruso, R., et al. (2014). *Methods in Molecular Biology, 1172*, 49–64.
Liu, G., Luais, E., & Gooding, J. J. (2011). *Langmuir, 27*, 4176–4183.
Love, J. C., Estroff, L. A., Kriebel, J. K., Nuzzo, R. G., & Whitesides, G. M. (2011). *Chemical Reviews, 105*, 1103–1169.
Mesnage, A., Lefevre, X., Jegou, P., Deniau, G., & Palacin, S. (2012). *Langmuir, 28*, 11767–11778.
Mrksich, M., & Whitesides, G. M. (1995). *Trends in Biotechnology, 13*, 228–235.
Nessark, F., Eissa, M., Baraket, A., Zine, N., Nessark, B., Zouaoui, A., et al. (2020). *Electroanalysis, 32*, 1795–1806.
Ott, C., Raicopol, M. D., Andronescu, C., Vasile, E., Hanganu, A., Pruna, A., et al. (2018). *Synthetic Metals, 235*, 20–28.
Pinson, J., & Podvorica, F. (2005). *Chemical Society Reviews, 34*, 429–439.
Polsky, R., Harper, J. C., Dirk, S. M., Arango, D. C., Wheeler, D. R., & Brozik, S. M. (2007). *Langmuir, 23*, 364–366.
Polsky, R., Harper, J.C., Wheeler, D.R., Arango, D.C., Brozik, S.M. (2008). *Angewandte Chemie* (International ed. in English), *47*, 2631–2634.
Polsky, R., Harper, J. C., Wheeler, D. R., Dirk, S. M., Arango, D. C., & Brozik, S. M. (2008). *Biosensors and Bioelectronics, 23*, 757–764.
Radi, A.-E., Muñoz-Berbel, X., Cortina-Puig, M., & Marty, J.-L. (2009). *Electrochimica Acta, 54*, 2180–2184.
Renault, J. P., Bernard, A., Bietsch, A., Michel, B., Bosshard, H. R., Delamarche, E., et al. (2003). *The Journal of Physical Chemistry B, 107*, 703–711.
Stine, R., Cole, C. L., Ainslie, K. M., Mulvaney, S. P., & Whitman, L. J. (2007). *Langmuir, 23*, 4400–4404.
Stewart, M. P., Maya, F., Kosynkin, D. V., Dirk, S. M., Stapleton, J. J., McGuiness, C. L., et al. (2004). *Journal of the American Chemical Society, 126*, 370.
Squillace, O., Esnault, C., Pilard, J.-F., & Brotons, G. (2017). *ACS Applied Materials & Interfaces, 9*, 42313–42326.
Tahara, K., Ishikawa, T., Hirsch, B. E., Kubo, Y., Brown, A., Eyley, S., et al. (2018). *Nano, 12*, 11520–11528.
Tlili, A., Jarboui, M. A., Abdelghani, A., Fathallah, D. M., & Maaref, M. A. (2005). *Materials Science and Engineering C, 25*, 490–495.
Tripoliti, E. E., Ioannidou, P., Toumpaniaris, P., Rammos, A., Pacitto, D., Lourme, J.-C., et al. (2020). *IEEE Reviews in Biomedical Engineering, 13*, 17–31.
Turner, A. P. F. (2000). *Science, 290*, 1315–1317.
Xia, Y., & Whitesides, G. M. (1998). *Angewandte Chemie, 37*, 550–575.
Yáñez-Sedeño, P., Campuzano, S., & Pingarrón, J. M. (2018). *Sensors, 18*, 675.
Zamfir, L.-G., Geana, I., Bourigua, S., Rotariu, L., Bala, C., Errachid, A., et al. (2011). *Sensors and Actuators, B, 159*, 178–184.
Zhang, D., & Wang, Y. (2006). *Materials Science and Engineering B, 134*, 9–19.
Zhang, F., Sautter, K., Larsen, A. M., Findley, D. A., Davis, R. C., Samha, H., et al. (2010). *Langmuir, 26*, 14648–14654.

# Review on Recent Advances in Urinary Biomarkers Based Electrochemical Sensors for Prostate Cancer Detection

**Meriem Mokni, Najla Fourati, Chouki Zerrouki, Ali Othmane, Asma Omezzine, and Ali Bouslama**

**Abstract** Prostate cancer (PCa) is the first most frequently diagnosed malignancy in man in Europe and is the third major case of males' cancer-related death. PCa screening and diagnosis are therefore societal and public health issues. Prostate specific antigen is the routine marker, but it is not specific for PCa. Several promising new biomarkers, including proteins, circulating tumor-derived DNA and RNA, and metabolites, are currently under clinical and analytical evaluation. The most promising ones are probably those present in urine, a valuable biological fluid that contains diverse biomarkers produced nearby by the prostatic tumour, and which can be easily collected with non-invasive sampling procedure. For each type of biomarker, there are already conventional assay techniques: namely ELISA and Western Blot for proteins and RT-PCR for DNA and RNA. Despite their undeniable metrological performances, these techniques remain expensive and require sophisticated equipment. Hence there is a need for ultra-sensitive, reliable and disposable tools for preclinical diagnosis of PCa. The most promising candidates are probably the electrochemical

M. Mokni · N. Fourati (✉) · C. Zerrouki
SATIE Laboratory, Cnam, Paris, France
e-mail: fourati@cnam.fr

M. Mokni
e-mail: mokni.meriem@hotmail.fr

C. Zerrouki
e-mail: zerrouki@cnam.fr

M. Mokni · A. Omezzine · A. Bouslama
LR12SP11 Laboratory, CHU Sahloul, Sousse, Tunisia
e-mail: asmaomezzine2@gmail.com

A. Bouslama
e-mail: ali.bouslama@rns.tn

A. Othmane
LIMA Laboratory, Faculté de Médecine de Monastir, Monastir, Tunisia
e-mail: ali.othmane54@gmail.com

 
O. Kanoun and N. Derbel (eds.), *Advanced Sensors for Biomedical Applications*,
Smart Sensors, Measurement and Instrumentation 38,
https://doi.org/10.1007/978-3-030-71225-9_8

biosensors. In this review are thus presented the recent advances in the design of electrochemical biosensors for the quantification of urinary biomarkers of prostate cancer.

**Keywords** Prostate cancer · Urinary biomarkers · Electrochemical sensors

## 1 Introduction

According to the World Health Organization (WHO), cancer is the second leading cause of death in the globally and accounted for one in six deaths in 2018 (World Health Organisation 2018). In Europe, the 2020 estimates reveal that the most diagnosed males' cancers are prostate, lung, colorectal, stomach and bladder (European Commission 2020). The diagnosis and prognosis of prostate cancer (PCa) are thus of prime importance. PCa is generally suspected when a mass is found in the prostate, after a digital rectal examination (DRE), and/or a prostate specific antigen (PSA) concentration in serum superior to the cut-off value of 10 ng/mL. Diagnosis confirmation is obtained mainly by the Gleason score after biopsy (Chistiakov et al. 2018). Although PCa is treatable at the early stages, it becomes castration resistant at the advanced stages and difficult to treat. PCa causes are not yet completely understood, but researchers have found several factors that might increase the risk of getting it: age, family history, ethnic origin, diet, obesity.... Hence, a rapid and sensitive diagnosis is essential before the cancer spread out to the other body organs.

Despite its use in routine clinical practice, PSA biomarker suffers from many drawbacks, mainly its low specificity and low positive predictive value as its concentration in patients sera, beyond the cut-off levels, may sometimes increases in benign prostate hyperplasia (BPH) or during prostate's manipulations (like digital rectal examination). PSA can thus only be used as a screening test, while the definitive diagnosis of prostate cancer requires a prostate biopsy (Duffy 2014).

Numerous novel diagnostic biomarkers are thus investigated in clinical uses to detect PCa in patients with a disease or abnormal condition. According to their molecular nature, these biomarkers can be roughly divided into three groups: DNA, RNA and protein biomarkers (Eskra et al. 2019). Based on their "localisation", the biomarkers can be found into tissue, serum, tumour cells or urine. Figure 1 gathers PCa current early diagnostic biomarkers and the corresponding assay techniques.

This large number of PCa biomarkers reflects the complex analytical and regulatory challenges for applying biomarkers in prostate cancer care. In this review, we have chosen to investigate only the urinary ones. In fact, compared to blood, where activation of proteases and generation of proteolytic breakdown products takes place within minutes of collection, urinary biomarkers are most stable and do not undergo significant proteolysis within several hours of collection.

Moreover, urinary proteomics presents an attractive approach to cancer biomarker discovery, not only for kidney/urological malignancies, but also for other systemic

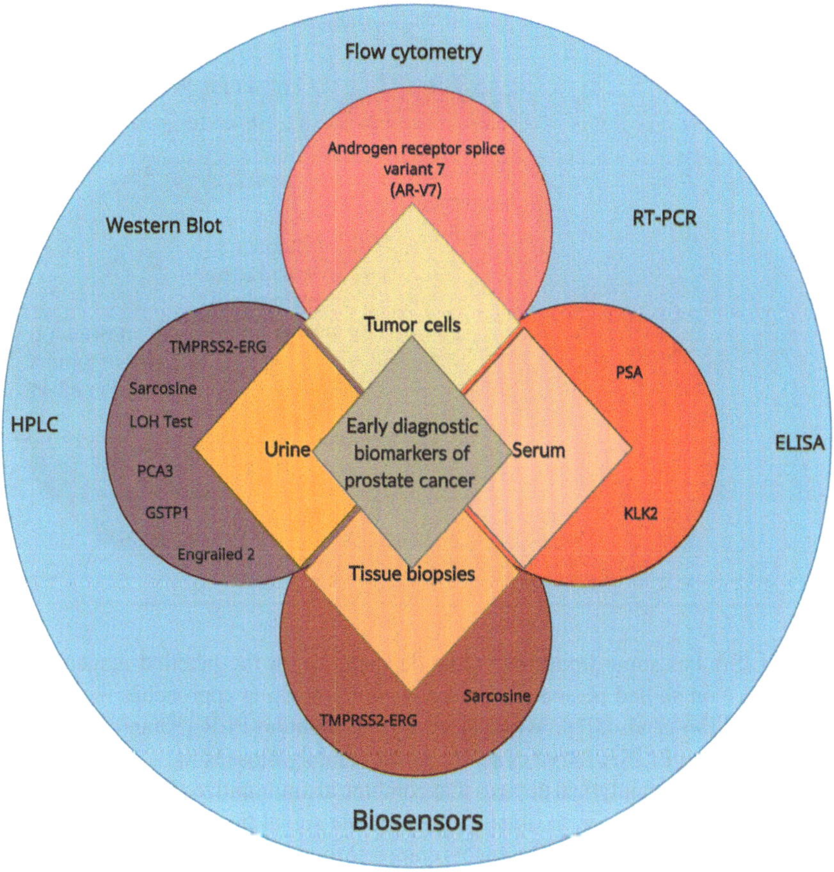


**Fig. 1** Current early diagnostic biomarkers of PCa (Rice and Stoyanova 2018; Qu et al. 2014)

malignancies. They also offer a great chance for the development of novel, non-invasive assays for the diagnosis, monitoring and the accurate diagnosis of PCa as illustrated in Fig. 2.

Many advantages favour the use of urine, as a detection medium for cancer biomarkers, over blood and tissues samples. First, urines are non-infectious for HIV and less infectious for many other pathogens. They have also the advantages of being readily available in large quantities using non-invasive collection procedures with a relative stability (Dhondt et al. 2018; Truong et al. 2018). And finally, as the prostate is connected to the urethra, all molecular changes especially caused by tumour invasion and reorganization of the extracellular matrix are reflected in urine (Truong et al. 2018).

Several highly sensitive and selective instruments are involved in prostate cancer diagnostic and prognostic. Enzyme-linked immunosorption assay (ELISA), which relies on fluorescent labelling, is considered as the golden standard method. How-

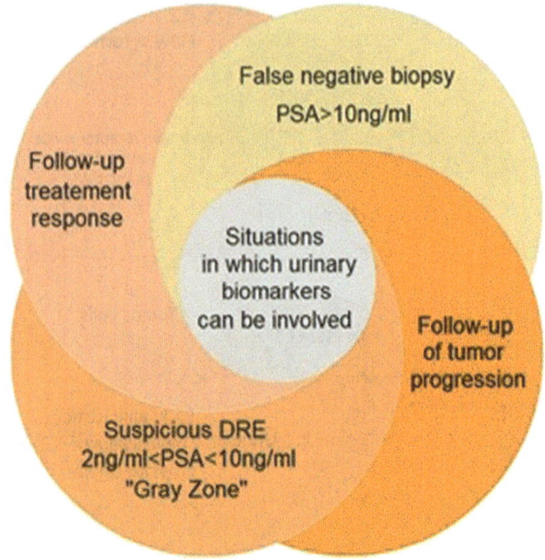


**Fig. 2** Clinical implications of urinary biomarkers

ever, ELISA has some limitations, like dependence on the labelled reagents, and the reliance on skilled personnel. Emerging and promising approaches include real time PCR (Tao et al. 2018; Wang et al. 2017), immuno-PCR (Khan et al. 2016), electrochemical ELISA (Arya et al. 2018), and mainly biosensors.

These latter are analytical devices that combine an immobilized biological sensing material and a transducer, to obtain a quantifiable signal that is proportional to the concentration of the analyte. Biosensors comprise three distinct components: (i) a bio-recognition element to recognize and bind the analyte of interest; (ii) a transducing element that converts the interactions of biomolecules into a quantifiable signal; and (iii) a readout system. Several studies have reported the design of biosensors for accurate, sensitive and early detection of cancers (Wang et al. 2006; Cheng et al. 2019; Cui et al. 2019), cancer metastasis monitoring (Cui et al. 2019; Quinchia et al. 2020) and further analyse effectiveness of anticancer chemotherapy drugs (Wang et al. 2006).

Among the large variety of available biosensors, the electrochemical ones are probably the most common ones, due to their relative ease of use, portability, high sensitivity, accuracy, and reliability. They have therefore received increasing attention, and efforts are currently devoted to the design of ultrasensitive electrochemical biosensors for cancer biomarkers detection.

Hereafter, an exhaustive summary to date on the electrochemical biosensors designed for the detection of the urinary biomarkers of prostate cancer. We will present the potential of each biomarker, prior to the functionalization strategy and further metrological performances of the developed biosensors. Future challenges in using electrochemical sensors for point-of-care testing are presented.

## 2 Prostate Cancer Antigen: PCA3

Prostate cancer antigen (PCA3) is a prostate specific non-protein coding RNA, that is highly expressed in prostate cancer cells and that is interestingly absent in non-cancer prostate pathologies, such as BPH and atypical prostatitis (Chistiakov et al. 2018; Whitman et al. 2008). Despite its moderate sensitivity (66.27%) and specificity of 80.94% (Wang et al. 2017), PCA3 test in urines, after DRE, can be considered as a promising non-invasive screening test and has the potential to provide valuable prognostic information (Whitman et al. 2008; Lee et al. 2020). Literature has reported in fact that PCA3 can be used to predict tumours' volumes and cancers stages and that it can be a strong predictor of extracapsular extension when associated with Gleason score (Truong et al. 2018; Whitman et al. 2008; Lee et al. 2020; Nakanishi et al. 2008).

Fort all these reasons, the Food and Drug Administration (FDA) approved the use of this biomarker for prostate cancer detection. In 2006, the Council of Europe approved the PROGENSA PCA3 urine test, which measures the PCA3 mRNA level normalized to that of PSA mRNA in urine. Six years later, the US FDA administration takes the same decision and approved PROGENSA PCA3 urine test (Truong et al. 2018; Lee et al. 2020).

Despite its great interest, only two studies reported on electrochemical biosensors for PCA3 detection. Soares et al. (2019) designed an impedance genosensor by functionalizing the surfaces of gold interdigital electrodes with a film of chitosan and multi-walled carbon nanotubes (MWCNT). They activated the carboxylic groups of the MWCNT by amidation, using the couple N-ethyl-N-(3-(dimethylamino)propyl) carbodiimide/N-hydroxysuccinimide (EDC/NHS: 1/1, M/M. Then, the formed NHS ester reacts the NH2 ending group of the PCA3 probe (TTTTTTTCCCAGGGATCTCTGTGCTTCC with one NH2-ending group). In this study, the authors have tested "synthetic" PCA3 with concentrations ranging from 0.1 fM to 1 μM. The detection limit of the designed sensor was of order of 128 pM.

The second genosensor was designed by Rodrigues et al. (2021): layer-by-layer films, containing alternating layers of gold nanoparticles stabilized with chondroitin sulfate, were deposited on Dropsens printed carbon electrodes. The surfaces were then functionalized with a complementary DNA sequence layer (PCA3 probe). Two electrochemical techniques were investigated in this study: impedance spectroscopy and cyclic voltammetry. The highest sensitivity was reached with the former technique. The corresponding detection limit was of order of 83 pM, is slightly inferior to that of Soares et al. (2019).

Notice that for both designed sensors, the authors have not tested the detection of PCA3 in urine samples.

## 3 Sarcosine

Sarcosine (Fig. 3) is a non-proteinogenic amino acid generated by glycine-N-methyltransferase in its biochemical cycle. This metabolite increases in urine during the progression of prostate cancer and metastatic prostate, and is found in negligible or even zero concentration in urine of healthy persons (Cernei et al. 2013). However, its role in carcinogenesis remains yet unknown (Cernei et al. 2013).

Sarcosine test in urine samples has the advantage of differentiating the patients with BPH from those with newly diagnosed prostate cancer. It can thus be considered as a valuable biomarker for PCa with discriminative clinical significance (Yousefi et al. 2020).

The usual technique for sarcosine detection is typically the high performance chromatography (Jiang et al. 2010). However, this method requires expensive equipment and specific columns. Several electrochemical biosensors have thus been designed in the last five years to detect sarcosine in urine.

Hroncekova et al. (2020) designed an enzymatic biosensor by immobilizing sarcosine oxydase on glassy carbon electrodes functionalized with Ti3C2TX MXene/chitosan nanocomposite. The hydrogen peroxide generated by the enzymatic reaction, in the presence of sarcosine, was detected by amperometry. The limit of detection of this biosensor was equal to 91.4 nM.

Using the same strategy, Li et al. (2019) have reported the design of an amperometric biosensor. The glassy carbon electrode was first modified by polyamic acid (PAA) and then by the glutaraldehyde before sarcosine oxidase (SOX) immobilisation. This sensor conception, allowing the sarcosine oxidation and then the peroxide liberation, was successfully used for the sarcosine detection in urine samples with a LOD value of 0.4 nM.

To improve the stability and the selectivity of their sensor, Rebelo et al. (2014) modified the carbon screen printed electrodes surfaces with a layer of metallic or semiconductor nanoparticles before, or, during SOX immobilization. The authors detected sarcosine in urine samples with a LOD value superior to those of the studies cited above.

Consequently, the process of the electrode surface functionalization has a key role in the design of sensitive sensors. The technology of molecular imprinting is increasingly investigated in the modification of the working electrodes' surfaces of electrochemical sensors. It consists in creating imprints in a highly cross-linked polymeric matrix. In general, the molecule of interest (template), adequate functional monomers and a crosslinker are pre-polymerized together, during a fixed duration and

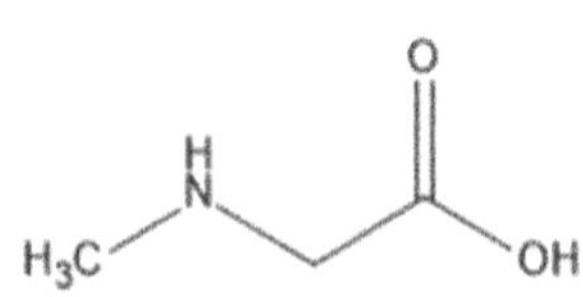


**Fig. 3** Chemical structure of sarcosine

under optimized physicochemical parameters (pH, ionic strength, solvent, supporting electrolyte, $\cdots$). Subsequently, bulk, or electrochemical polymerization process will lead to the formation of polymeric chains which self-organize around the templates through covalent, hydrogen or Van Der Waals interactions. Then, the templates are removed from the vicinity the polymeric matrix, by using an extractor agent, leaving imprints, with a complementary structure in terms of geometrical shape and chemical functions, which are able to recognise the analytes of interest (Mattiasson and Ye 2015; Haupt et al. 2018).

Concerning Sarcosine detection, Tang et al. (2020) used Magnetic $Fe_3O_4$ nanoparticles embedded zeolitic imidazolate framework-8 (ZIF-8) and functionalized them with a molecularly imprinted polymer (MIP). The synthesized magnetic $Fe_3O_4$@ZIF-8@MIP was self-assembled onto gold electrodes and used as the sensing unit of the electrochemical sensor. The authors investigated the cyclic voltammetry technique to monitor the binding of sarcosine to the MIP and the further tests in urine samples. The designed sensor has a limit of detection of 0.4 pM, largely inferior to those obtained with the enzymatic modification (Hroncekova et al. 2020; Li et al. 2019; Rebelo et al. 2014). The MIP based sensor sensitivity was equal 0.708 μA/pM.

Özkütük et al. (2016) designed a MIP based potentiometric sensor for sarcosine monitoring. Methacryloylamido histidine and ethylene glycol dimethacrylate were the functional monomer and the cross-linking agent respectively. The MIP was synthesized by emulsion polymerization. The designed sensor was selective of sarcosine and its LOD was of order of 135 pM.

## 4 Engrailed-2 Protein

Engrailed-2 (EN2) is a transcriptional repressor and a member of group of homeodomain-containing transcription factors called HOX genes that determine the early identification of cells and tissues in initial embryonic development involving transcriptional and translational regulation. The dysregulation of these genes occurs in most common cancer (Marszall et al. 2015; Challacombe et al. 2013).

EN2 is expressed in prostate cancer and is secreted in urine but it is not found in normal gland, making it a good potential diagnostic marker with a sensitivity of 66.7% and a specificity of 88.2% (Morgan et al. 2011). Besides, Pandha et al. reported that EN2 urinary levels reflected the volume and stage of the cancerous tumour (Pandha et al. 2012). Interestingly, prior DRE is not required unlike for PCA3 test, which improves the acceptability to the patients (Marszall et al. 2015).

EN2 essays for clinical PCa studies are using a typical enzyme-linked immunosorbent assay (ELISA) (Marszall et al. 2015; Morgan et al. 2011; Pandha et al. 2012; Do Carmo Silva et al. 2017; Morgan et al. 2013; Killick et al. 2013; McGrath et al. 2013). As for EN2, only two electrochemical biosensors were reported in the literature. Prior to the fabrication of their sensor, Lee et al. (2015) designed a DNA probe which specifically recognizes the EN2 protein. The authors used, as a starting point, the DNA sequence (5'-TAATTA-3') well known for its ability to strongly bind to EN2, and optimized then the affinity by modifying the extra sequences of the specific binding sites. The impedimetric biosensor was elaborated by grafting the DNA probes on gold electrodes modified by gold nanoparticles to increase the sensitivity. The obtained sensor was tested in artificial biological media, and the limit of detection was of order of 5.62 fM. It was however not been tested and validated on urines from patients. Settu et al. (2017), used the same (5'-TAATTA-3') to elaborate a cyclic voltammetry biosensor using a screen-printed carbon-graphene electrode. The principle is based on the activation of the surface of the electrode, rich in carboxylic groups, by the EDC/NHS couple in order to covalently graft the aptamer. The announced limit of detection was of 38.5 nM, largely superior to that of (Lee et al. 2015).

## 5 Micro-RNA (miRNA)

MiRNAs are described as small, 20–22 nucleotides non-coding RNAs that regulate the expression of a multitude of target genes through binding to the 3'-untranslated region (3'-UTR) of their target mRNAs (Zhou et al. 2017). It was reported that over than 17 microRNA types are involved in prostate cancer and are known for their role as oncogenes, tumour suppressors and metastasis regulators (Kanwal et al. 2017). MiRNAs can thus be used as prognostic markers as they are commonly upregulated in prostate cancer (Mohammadi et al. 2019).

The most frequent methods for the quantification and identification of microRNAs are based on molecular biology techniques such as polymerase chain reaction (PCR), DNA microarray, and Northern blot techniques (Mohammadi et al. 2019).

Numerous electrochemical genosensors have been designed for microRNA detection. They have the benefit of being sensitive enough to allow skipping the extraction steps of microRNA from the body fluids and amplification steps by conventional PCR (Mohammadi et al. 2019). However, most of them concerned the detection of microRNA in blood samples (Mohammadi et al. 2019).

To the best of our knowledge, only one publication reported the conception of electrochemical biosensor for the detection of miRNAs in urine. Smith et al. (2017) designed an impedimetric biosensor, based on a modified glassy carbon electrode (CGE), for the detection of miR21. The CGE was modified by sulfonic acid and subsequent chlorination prior to DNA probes grafting. The recognition was achieved by hybridization between the DNA probes and micro RNA. The biosensor was selective, as it was able to discriminate between miR-21, three point-mutated miR-21 sequences, and miR-16. Its limit of detection was of order of 20 fM.

## 6 $\alpha$-Methylacyl-CoA Racemase (AMACR)

Alpha-methyl CoA-racemase (AMACR) is an enzyme involved in the catabolism of bile acid metabolites and the $\beta$-oxidation of branched fatty acids (Ji and Chen 2019). The literature reported an over expression of AMACR protein in cancerous tissues and their low expression in the benign ones (Jiang et al. 2013). Ouyang et al. (2018) demonstrated that an AMACR test in urine sediments has a sensitivity of 70% and a specificity of 71% and that a combine use of AMACR and PCA3 scores in a dual-marker test increased sensitivity to 81% and specificity to 84%. Besides, Ji and Chen (2019) demonstrated that the diagnostic model combining serum PSA and urine AMACR has a better diagnostic value than PSA alone especially grey-zone PSA patients and could reduce unnecessary biopsy.

AMACR can also be used as a follow-up tool for hormonal therapy. Suzue et al. (2005) have carried out a study on patients with localized prostate cancer and who were under hormonal treatment. They showed a significant decrease in AMACR level expression. Further studies are however needed to explore correlation between AMACR level and the treatment efficiency.

Concerning the field of biosensors, Wang and Yau (2014) designed, in 2014, an immunosensor for AMACR detection in serum and urine, with a sensitivity of 0.0487 mA.mL/ng and 12.9 mA.mL/ng respectively. They used a sandwich structure to detect the enzyme: a primary capture antibody grafted on the sensing area via a polyaniline film and a secondary detection antibody conjugated to horseradish peroxidase (HRP). The particularity of this biosensor is the use of a gating electrode and applying a gating voltage to the immune complex to provide signal amplification. The detection limit of the designed sensor was of order of 100 ng/ml.

## 7 Microseminoprotein-Beta (MSMB)

Microseminoprotein-beta (MSMB) is a small non-glycosylated peptide (94 amino acids) (Rebelo et al. 2016) that regulates cell growth by promoting apoptosis in normal healthy prostate while in malignancy, the opposite phenomenon is observed leading to uncontrolled growth of cells (Shrivastava et al. 2020). According to several studies carried out by ELISA measurements on serum and urines, MSMB is reduced in PCa in proportion to the tumour stage (Sjöblom et al. 2016).

To the best of our knowledge, only one electrochemical sensor has been reported in the literature. Rebelo et al. (2016) designed a sensitive impedimetric sensor to detect MSBM in urine and serum. Two types of MIPs were constructed. The first one was obtained by the electropolymerisation of caffeic acid with MSBM in phosphate buffer solution. The second one was constructed in the same conditions as the first, but charged dopamine was added to increase the binding site label. The templates were removed from the vicinity of the polymeric matrix by incubating the electrodes overnight in a $H_2SO_4$ solution (1 mol/l) at 45 °C. Tested in urine synthetic solution, the two MIPs based sensors presented a limit of detection of 0.18 ng/ml and 0.079 ng/ml respectively. Their corresponding sensitivities, per decade of urine concentration, were of order of 6.38 μA and 13.52 μA respectively.

## 8 TMPRSS2: ERG Transcripts

Chromosomal rearrangements seem to occur in prostate cancer, and specifically for chromosome 21 which juxtaposes the 5' untranslated region of TMPRSS2 to the ERG oncogene and is observed in 40–80% of prostate cancers (Dhondt et al. 2018; Sanda et al. 2017). The TMPRSS2 expression is regulated by androgens and codes for the transmembrane serine protease 2 highly expressed by normal and cancerous prostate cells. The gene ERG encodes transcription factors involved in regulating cell growth, cell differentiation and carcinogenesis (Durand et al. 2010). The ERG gene is activated by fusion with TMPRSS2 and might be responsible of tumour progression and stimulation of the overexpression of transcription factors which could lead to epigenetic reprogramming and a dysregulation of the pathways of apoptosis (Durand et al. 2010). Diverse prospective observational studies highlighted the prognostic value of the TMPRSS2:ERG transcripts in PCa (Sanda et al. 2017; Tomlins et al. 2016; Leyten et al. 2013). In fact, it has been demonstrated that TMPRSS2:ERG transcript assay in urine after DRE adds important value in the prediction of clinical tumour stage as well as biopsy Gleason score and extracapsular extension of the tumour. Besides, combining urinary testing of PCA3 and TMPRSS2:ERG transcripts allows to avoid unnecessary biopsy as they have together a robust sensitivity for detecting aggressive PCa (Sanda et al. 2017; Tomlins et al. 2016).

There is no work published until now about development of a direct electrochemical genosensors for the fusion gene assay in urine. However, Koo et al. reported an amperometric readout strategy after amplification of RNA extracted from urine samples by reverse transcription recombinase polymerase amplification RT-RPA (Koo et al. 2019).

## 9 Conclusion and Perspectives

Because of the clinical heterogeneity of the disease and the need to differentiate cancer from inflammatory and benign conditions, using a single biomarker for a robust clinical screening test for PCa is certainly not sufficient. Researchers have thus begun testing the efficiency of assays combining sera and urinary biomarkers. Urine is in fact a highly desirable biomarker medium as it can be collected by non-invasive techniques.

This review highlighted the fact that there are many urinary biomarkers with a large potential for PCa screening, but, today, none of them has reached a widespread use and the majority of them needs to be validated in real clinical and analytical conditions to fulfil the European and American criteria for clinical use.

In contrast to its relative ease and robustness, the use of electrochemistry, as a tool for the investigation of urinary biomarkers is yet limited, as shown by the small number of publications related to this subject, but we are convinced of the great potential offered by these devices, and of the tremendous contribution they could bring to oncologists, like what the glucometer, has bring to the medical community.

Besides, the molecular imprinted polymer approaches associated to electrochemical transduction, offer a very promising way to design MIP-based biosensors directed towards simultaneous correlated detections of several pertinent biomarkers.

## References


Arya, S. K., & Estrela, P. (2018). Recent advances in enhancement strategies for electrochemical ELISA-based immunoassays for cancer biomarker detection. *Sensors*, *18*(7), 2010.

Cernei, N., Heger, Z., Gumulec, J., Zitka, O., Masarik, M., Babula, P., et al. (2013). Sarcosine as a potential prostate cancer biomarker-A review. *International Journal of Molecular Sciences*, *14*(7), 13893–13908.

Challacombe, B., Fitzpatrick, J., & Kirby, R. (2013). PCA3 and other urinary markers. In J.S. Jones (ed.), *Prostate cancer diagnosis. PSA, biopsy and beyond* (pp. 61–71.). Totowa: Humana Press.

Cheng, N., Du, D., Wang, X., Liu, D., Xu, W., Luo, Y., et al. (2019). Recent advances in biosensors for detecting cancer-derived exosomes. *Trends in Biotechnology*, *37*(11), 1236–1254.

Chistiakov, D.A., Myasoedova, V.A., Grechko, A.V., Melnichenko, A.A., & Orekhov, A.N. (2018). New biomarkers for diagnosis and prognosis of localized prostate cancer. In *Seminars in Cancer Biology 2018* (pp. 9–16). Elsevier.

Cui, F., Zhou, Z., & Zhou, H. S. (2019). Measurement and analysis of cancer biomarkers based on electrochemical biosensors. *Journal of The Electrochemical Society*, *167*(3), 037525.

Dhondt, B., Van Deun, J., Vermaerke, S., de Marco, A., Lumen, N., De Wever, O., et al. (2018). Urinary extracellular vesicle biomarkers in urological cancers: From discovery towards clinical implementation. *The International Journal of Biochemistry & Cell Biology*, *99*, 236–256.

Duffy, M. J. (2014). PSA in screening for prostate cancer: More good than harm or more harm than good? *Advances in Clinical Chemistry*, *66*, 1–23.

Durand, X., Xylinas, E., Ploussard, G., & De la Taille, A. (2010). Biomarqueurs urinaires du cancer de prostate. Progrès en urologie, *20*(13), 1184–1191.

Eskra, J. N., Rabizadeh, D., Pavlovich, C. P., Catalona, W. J., & Luo, J. (2019). Approaches to urinary detection of prostate cancer. *Prostate Cancer and Prostatic Diseases*, *22*(3), 362–381.

European Commission. (2020). 2020 Cancer incidence and mortality in EU-27 countries. https://ec.europa.eu/jrc/en/news/2020-cancer-incidence-and-mortality-eu-27-countries.

Haupt, K., Medina Rangel, P. X., & Bui, B. T. S. (2020). Molecularly imprinted polymers: Antibody mimics for bioimaging and therapy. *Chemical Reviews*, *120*(17), 9554–9582.

Hroncekova, S., Bertok, T., Hires, M., Jane, E., Lorencova, L., Vikartovska, A., et al. (2020). Ultrasensitive Ti3C2TX MXene/Chitosan nanocomposite-based amperometric biosensor for detection of potential prostate cancer marker in urine samples. *Processes*, *8*(5), 580.

Ji, J., & Chen, X. (2019). Prostate cancer diagnosis using urine sediment analysis-based $\alpha$-Methylacyl-CoA racemase score: A single-center experience. *26*(1), 1073274819887697.

Jiang, N., Zhu, S., Chen, J., Niu, Y., & Zhou, L. (2013). A-methylacyl-CoA racemase (AMACR) and prostate-cancer risk: A meta-analysis of 4,385 participants. *PLOS One*, *8*(10), e74386.

Jiang, Y., Cheng, X., Wang, C., & Ma, Y. (2010). Quantitative determination of sarcosine and related compounds in urinary samples by liquid chromatography with tandem mass spectrometry. *Analytical Chemistry*, *82*(21), 9022–9027.

Kanwal, R., Plaga, A. R., Liu, X., Shukla, G. C., & Gupta, S. (2017). MicroRNAs in prostate cancer: Functional role as biomarkers. *Cancer Letters*, *407*, 9–20.

Khan, A. H., & Sadroddiny, E. (2016). Application of immuno-PCR for the detection of early stage cancer. *Molecular and Cellular Probes*, *30*(2), 106–112.

Killick, E., Morgan, R., Launchbury, F., Bancroft, E., Page, E., Castro, E., et al. (2013). Role of engrailed-2 (EN2) as a prostate cancer detection biomarker in genetically high risk men. *Scientific Reports*, *3*, 2059.

Koo, K.M., Wee, E.J.H., & Trau, M. (2019). *Advancing gene fusion detection towards personalized cancer nanodiagnostics*. Basingstoke: Springer Nature.

Lee, D., Shim, S.R., Ahn, S.T., Oh, M.M., Moon, D.G., Park, H.S., et al. (2020). Diagnostic performance of the prostate cancer antigen 3 test in prostate cancer: A systematic review and meta-analysis. *Clinical Genitourinary Cancer*.

Lee, S., Jo, H., Her, J., Lee, H. Y., & Ban, C. (2015). Ultrasensitive electrochemical detection of engrailed-2 based on homeodomain-specific DNA probe recognition for the diagnosis of prostate cancer. *Biosensors and Bioelectronics*, *66*, 32–38.

Leyten, G. H. J. M., Wierenga, E. A., Michiel Sedelaar, J. P., van Oort, I. M., Futterer, J. J., Barentsz, J. O., et al. (2013). Value of PCA3 to predict biopsy outcome and its potential role in selecting patients for multiparametric MRI. *International Journal of Molecular Sciences*, *14*(6), 11347–11355.

Li, J., Ma, J., Zhang, Y., Zhang, Z., & He, G. (2019). An amperometric biosensor for the assay of sarcosine based on the cross coupled chemical and electrochemical reactions with practical applications. *Journal of Electroanalytical Chemistry*, *833*, 568–572.

Marszall, M. P., Sroka, W., Adamowski, M., Slupski, P., Jarzemski, P., Siodmiak, J., et al. (2015). Engrailed-2 protein as a potential urinary prostate cancer biomarker: A comparison study before and after digital rectal examination. *European Journal of Cancer Prevention*, *24*(1), 51–6.

Mattiasson, B. & Ye, L. (2015). *Molecularly imprinted polymers in biotechnology* (vol. 150). Berlin: Springer.

McGrath, S. E., Michael, A., Morgan, R., & Pandha, H. (2013). EN2: A novel prostate cancer biomarker. *Biomarkers in Medicine*, *7*(6), 893–901.

Mohammadi, H., Yammouri, G., & Amine, A. (2019). Current advances in electrochemical genosensors for detecting microRNA cancer markers. *Current Opinion in Electrochemistry*, *16*, 96–105.

Morgan, R., Boxall, A., Bhatt, A., Bailey, M., Hindley, R., Langley, S., et al. (2011). Engrailed-2 (EN2): A tumor specific urinary biomarker for the early diagnosis of prostate cancer. *Clinical Cancer Research: An Official Journal of the American Association for Cancer Research*, *17*(5), 1090–1098.

Morgan, R., Bryan, R. T., Javed, S., Launchbury, F., Zeegers, M. P., Cheng, K. K., et al. (2013). Expression of engrailed-2 (EN2) protein in bladder cancer and its potential utility as a urinary diagnostic biomarker. *European Journal of Cancer*, *49*(9), 2214–2222.

Nakanishi, H., Groskopf, J., Fritsche, H. A., Bhadkamkar, V., Blase, A., Kumar, S. V., et al. (2008). PCA3 molecular urine assay correlates with prostate cancer tumor volume: Implication in selecting candidates for active surveillance. *The Journal of Urology*, *179*(5), 1804–1810.

Ouyang, B., Bracken, B., Burke, B., Chung, E., Liang, J., Ho, S.M. (2009). A duplex quantitative polymerase chain reaction assay based on quantification of alpha-methylacyl-CoA racemase transcripts and prostate cancer antigen 3 in urine sediments improved diagnostic accuracy for prostate cancer. The Journal of Urology **181**(6), 2508–2513. Discussion 2513–2504.

Özkütük, E. B., Diltemiz, S. E., Avcı, Ş., Uğurağ, D., Aykanat, R. B., Ersöz, A., et al. (2016). Potentiometric sensor fabrication having 2D sarcosine memories and analytical features. *Materials Science and Engineering: C*, *69*, 231–235.

Pandha, H., Sorensen, K. D., Orntoft, T. F., Langley, S., Hoyer, S., Borre, M., et al. (2012). Urinary engrailed-2 (EN2) levels predict tumour volume in men undergoing radical prostatectomy for prostate cancer. *BJU International*, *110*(6b), E287–E292.
Qu, M., Ren, S. C., & Sun, Y. H. (2014). Current early diagnostic biomarkers of prostate cancer. *Asian Journal of Andrology*, *16*(4), 549–554.
Quinchia, J., Echeverri, D., Cruz-Pacheco, A. F., Maldonado, M. E., & Orozco, J. (2020). Electrochemical biosensors for determination of colorectal tumor biomarkers. *Micromachines*, *11*(4), 411.
Rebelo, T. S., Pereira, C. M., Sales, M. G., Noronha, J. P., Costa-Rodrigues, J., Silva, F., et al. (2014). Sarcosine oxidase composite screen-printed electrode for sarcosine determination in biological samples. *Analytica Chimica Acta*, *850*, 26–32.
Rebelo, T. S., Pereira, C. M., Sales, M. G. F., Noronha, J., & Silva, F. (2016). Protein imprinted materials designed with charged binding sites on screen-printed electrode for microseminoprotein-beta determination in biological samples. *Sensors and Actuators B: Chemical*, *223*, 846–852.
Rice, M. A., & Stoyanova, T. (2018). Biomarkers for diagnosis and prognosis of prostate cancer. *Prostatectomy* (IntechOpen).
Rodrigues, V. C., Soares, J. C., Soares, A. C., Braz, D. C., Melendez, M. E., Ribas, L. C., et al. (2021). Electrochemical and optical detection and machine learning applied to images of genosensors for diagnosis of prostate cancer with the biomarker PCA3. *Talanta*, *222*, 121444.
Sanda, M. G., Feng, Z., Howard, D. H., Tomlins, S. A., Sokoll, L. J., Chan, D. W., et al. (2017). Association between combined TMPRSS2: ERG and PCA3 RNA urinary testing and detection of aggressive prostate cancer. *JAMA Oncology*, *3*(8), 1085–1093.
Settu, K., Liu, J.-T., Chen, C.-J., & Tsai, J.-Z. (2017). Development of carbon?graphene-based aptamer biosensor for EN2 protein detection. *Analytical Biochemistry*, *534*, 99–107.
Shrivastava, P., Garg, H., Bhat, M., Dinda, A., & Kumar, R. (2020). Urinary prostate-specific antigen and microseminoprotein-beta levels in men with and without prostate cancer: A prospective cohort study. *Indian Journal of Urology: IJU: Journal of the Urological Society of India*, *36*(1), 50–55.
Silva, J. D. C., Vesely, S., Novak, V., Luksanova, H., Prusa, R., & Babjuk, M. (2017). Engrailed 2 in urine fails as a marker in prostate cancer detection. *European Urology Supplements*, *16*(10), e2646.
Sjöblom, L., Saramäki, O., Annala, M., Leinonen, K., Nättinen, J., Tolonen, T., et al. (2016). Microseminoprotein-beta expression in different stages of prostate cancer. *PLOS One*, *11*(3).
Smith, D. A., Newbury, L. J., Drago, G., Bowen, T., & Redman, J. E. (2017). Electrochemical detection of urinary microRNAs via sulfonamide-bound antisense hybridisation. *Sensors and Actuators B: Chemical*, *253*, 335–341.
Soares, J. C., Soares, A. C., Rodrigues, V. C., Melendez, M. E., Santos, A. C., Faria, E. F., et al. (2019). Detection of the prostate cancer biomarker PCA3 with electrochemical and impedance-based biosensors. *ACS Applied Materials & Interfaces*, *11*(50), 46645–46650.
Suzue, K., Montag, A. G., Tretiakova, M., Yang, X. J., & Sahoo, S. (2005). Altered expression of $\alpha$-methylacyl-coenzyme A racemase in prostatic adenocarcinoma following hormone therapy. *American Journal of Clinical Pathology*, *123*(4), 553–561.
Tang, P., Wang, Y., & He, F. (2020). Electrochemical sensor based on super-magnetic metal-organic framework@ molecularly imprinted polymer for Sarcosine detection in urine. *Journal of Saudi Chemical Society*, *24*(8), 620–630.
Tao, Z., Shen, M., Zheng, Y., Mao, X., Chen, Z., Yin, Y., et al. (2010). PCA3 gene expression in prostate cancer tissue in a Chinese population: Quantification by real-time FQ-RT-PCR based on exon 3 of PCA3. *Experimental and Molecular Pathology*, *89*(1), 58–62.
Tomlins, S. A., Day, J. R., Lonigro, R. J., Hovelson, D. H., Siddiqui, J., Kunju, L. P., et al. (2016). Urine TMPRSS2: ERG plus PCA3 for individualized prostate cancer risk assessment. *European Urology*, *70*(1), 45–53.
Truong, M., Yang, B., & Jarrard, D. F. (2013). Toward the detection of prostate cancer in urine: A critical analysis. *The Journal of Urology*, *189*(2), 422–429.

Wang, J. (2006). Electrochemical biosensors: Towards point-of-care cancer diagnostics. *Biosensors and Bioelectronics*, *21*(10), 1887–1892.

Wang, J., & Yau, S. T. (2014). Detection of $\alpha$-methylacyl-CoA racemase in serum and urine using a highly sensitive electrochemical immunodetector. *Electroanalysis*, *26*(7), 1441–1444.

Wang, T., Qu, X., Jiang, J., Gao, P., Zhao, D., Lian, X., et al. (2017). Diagnostic significance of urinary long non-coding PCA3 RNA in prostate cancer. *Oncotarget*, *8*(35), 58577.

Whitman, E. J., Groskopf, J., Ali, A., Chen, Y., Blase, A., Furusato, B., et al. (2008). PCA3 score before radical prostatectomy predicts extracapsular extension and tumor volume. *The Journal of Urology*, *180*(5), 1975–1979.

World Health Organisation. (2018). Cancer, overview, prevention & management. https://www.who.int/health-topics/cancer#tab=tab_1.

Yousefi, M., Qujeq, D., Shafi, H., & Tilaki, K. H. (2020). Serum and urine levels of sarcosine in benign prostatic hyperplasia and newly diagnosed prostate cancer patients. *Journal of Kermanshah University of Medical Sciences*, *24*(1).

Zhou, Y.-J., Yang, H.-Q., Xia, W., Cui, L., Xu, R.-F., Lu, H., et al. (2017). Down-regulation of miR-605 promotes the proliferation and invasion of prostate cancer cells by up-regulating EN2. *Life Sciences*, *190*, 7–14.

# Recent Advances in Ultrasensitive miRNA Biomarkers Detection

**Khouloud Djebbi, Mohamed Bahri, Mohamed Amin Elaguech, Rong Tian, Shi Biao, Chaker Tlili, and Deqiang Wang**

**Abstract** Recently, microRNAs gain a great interest in the bio-molecular field due to their fundamental role in clinical diagnostics. In this chapter, we generally discussed the biogenesis of microRNAs and the progress made for their detection. Researchers have widely tried to investigate their effort to build a sensitive, selective, and accurate platform for microRNA detection. To date, multiple techniques have been developed, ranging from the old conventional method (northern blot, RT-PCR, microarrays) to the newly established ones (biosensors, nanopores). However, given the various challenges related to miRNA detection, such as low abundance, small size, and high level of sequence similarity, different enzymatic and non-enzymatic amplification approaches were successfully exploited to improve such devices' sensitivity. Among these strategies, HCR, RCA, nanomaterials, and the use of enzyme-based target recycling like DSN enzyme. On the other hand, the combination of different methods

K. Djebbi · M. Bahri · M. A. Elaguech · R. Tian · S. Biao · C. Tlili (✉) · D. Wang (✉)
Chongqing Key Laboratory of Multi-scale Manufacturing Technology, Chongqing, PR China
e-mail: chakertlili@cigit.ac.cn

D. Wang
e-mail: dqwang@cigit.ac.cn

K. Djebbi
e-mail: Khouloud@cigit.ac.cn

M. Bahri
e-mail: bahrimohamed@cigit.ac.cn

M. A. Elaguech
e-mail: mohamed@cigit.ac.cn

R. Tian
e-mail: tianrong@cigit.ac.cn

S. Biao
e-mail: shibiao@cigit.ac.cn

Institute of Green and Intelligent Technology, Chinese Academy of Sciences, Chongqing 400714, PR China

Chongqing School, University of Chinese Academy of Sciences, Chongqing 400714, PR China

 
O. Kanoun and N. Derbel (eds.), *Advanced Sensors for Biomedical Applications*, Smart Sensors, Measurement and Instrumentation 38,
https://doi.org/10.1007/978-3-030-71225-9_9

has emerged as an ideal option for further enhancement of the sensitivity. In the end, knowing that the expression of a single miRNA is not enough to identify one specific disease, it is usually necessary to implant a simultaneous and multiplexed technique for more sophisticated and efficient diagnostic tools.

**Keywords** MicroRNAs · Optical biosensing · Electrochemical biosensors · Field-effect transistor (FET) · Nanopores assay · Nanomaterials · Amplification signal · Enzymes

# 1 MicroRNAs Overview

Around 98% of the human genome stand for non-proteins coding DNA. These specific regions are transcribed to generate functional non-coding RNAs (ncRNAs). Among them, microRNAs one of the highly reviewed forms of ncRNAs. MicroRNAs (miRNAs) are defined as a class of small endogenous RNAs including about 18–24 nucleotides. The Ruvkun and Ambros are the first groups to discover microRNA (line-4) in 1993. Over since, this significant finding has revolutionized the molecular biology field (Feinbaum et al. 2004; Wightman and Ruvkun 1993). MicroRNAs are key features for RNA splicing and post-transcriptional regulation of gene expression. To date, tremendous effort has been made in this field to comprehend the biogenesis and the mechanism of miRNAs action.

## *1.1 MiRNA Biogenesis and Functions*

MiRNA biogenesis process goes through different steps. In the nucleus, RNA polymerase II generally binds to a promoter of the genome sequence transcribes miRNA genes, and so it generates the primiRNA with a length range from 1 to 3 kb. The resulting transcript is then split with the microprocessor complex Drosha DGGR8 (DiGeorge Syndrome Critical Region 8), releasing a stem-loop of about 70–100 nucleotides named pre-miRNA. The latter is carried via the aid of Exportin-5 to the cytoplasm, where the RNase-III enzyme Dicer will further cut it to obtain the 18–24 dsmiRNA. Afterward, the subsequent miRNA is hydrolyzed into single strand oligos. It yields the mature miRNA, which plays an essential and functional role in the RNA-induced silencing complex (RISC) with the argonaute (AGO) protein family. While the second miRNA strand termed "passenger" is frequently disintegrated or incorporated in downstream regulation effect along with the regulation of miRNA homeostasis. Generally, the 3'-untranslated region (3' UTR) represents the miRNA/mRNA binding sites using complementary Watson-Crick base pairings. Yet, miRNAs can interact with other regions, counting the 5' UTR, gene promoters, and coding sequences (Broughton et al. 2016). The RISC complex provokes whether degradation or translational repression or sequestration/ destabilization via

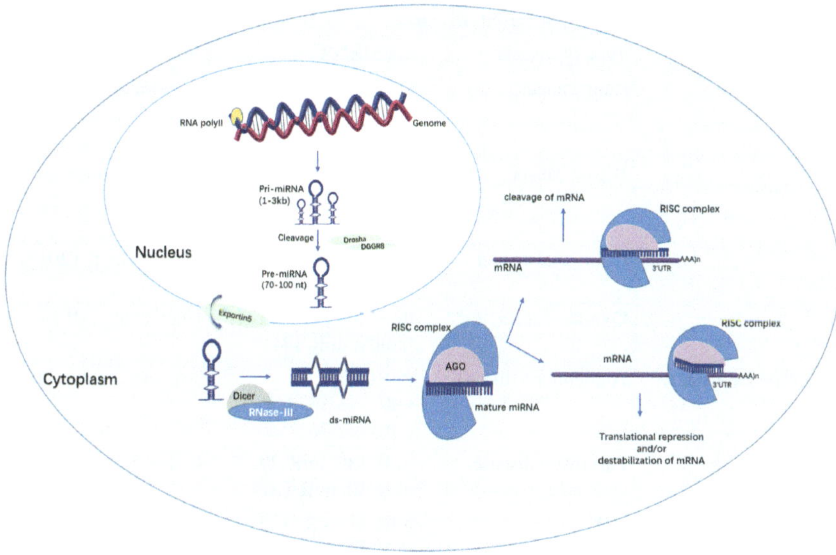


**Fig. 1** Overview of miRNA biogenesis and regulation

a shortage of mRNA's polyA tail from translational machinery. At the same time, the microRNA hybridizes with its complementary mRNA target. Definite evidence has pointed out that one unique miRNA might have multiple targets, and similarly, a single gene can be regulated by numerous miRNAs. The degree of complementarity between miRNA and mRNA denotes a vital feature of the regulatory mechanism. A superior level of complementarity facilitates the mRNA degradation due to its cleavage mechanism process. On the other hand, a mismatch induces the translational repression mechanism (Fig. 1). It is ascertained that the human genome encodes for more than 1000 miRNAs, which regulate in turn approximately 60% of gene expression in different biological activities (Yu and Pan 2012). The various processes in which miRNAs are involved include cell proliferation, apoptosis, immunity, neuronal patterning, hematopoietic differentiation, and fat metabolism,

## 1.2 *MicroRNAs in Different Pathological Processes*

MicroRNAs give insight into the etiology and the progression of several pathologies. The expression profile of miRNA represents a core characteristic of all human diseases. It is a double-edged sword. It can provoke the disease or suppress it since some of these microRNAs are overexpressed in cells, tissues, VOC (volatile organic compounds), vesicles like exosomes, and biological fluids (blood, saliva, cerebrospinal fluid), while others are expressed at lower levels. For example, miRNA-21 is considered to be overexpressed in almost 80% of the tumor samples, while miRNA-143,

**Table 1** Some miRNA involved in various diseases

| Disease | Type of disease | microRNAs | Refs. |
|---|---|---|---|
| Cancer | Lung cancer | miR-155, miR-21, miR-126, Let7, miR-210 | Wu et al. (2019) |
| | Breast cancer | miR-21, miR-16, miR-17, miR-106a, miR-126 | Singh and Mo (2013) |
| | Colorectal cancer | miR-17-3p, miR-141, miR-486-3p, miR-601 | Schetter et al. (2012) |
| | Ovarian cancer | miR-200 family, Let7 family, miR-141 | Chen et al. (2019) |
| Neurological disease | Alzheimer's disease | miR-127-3p, miR-34a-5p, miR-93, miR-342-3p | Angelucci et al. (2019) |
| | Parkinson disease | miR-485, miR-29, miR-30, miR-Let7 | Goh et al. (2019) |
| Infectious disease | HIV | miR-21,miR -122, miR-223, miR-3162-3p | Tribolet et al. (2020) |
| | VSV | miR-197, miR-629, miR-363, miR-132, miR-122 | |
| | Malaria | miR-15a-3p, miR-30c-5p, miR-30e-5p | |
| Diabetes | Insulin resistance | miR-195, miR-135a, miR-7, miR-499, miR-96 | Feng et al. (2016) |
| | Type 2 diabetes | miR-1249, miR-320b, miR-572, miR-126 | |
| Vascular disease | Angiogenesis | miR-221, miR-222, miR-210, miR-130a, Let7 | Qin and Zhang (2011) |
| | Atherosclerosis and restenosis | miR-125a,b, Let7, miR214, miR-146, miR-145 | |

miRNA-126, and miRNA-145 are downregulated in approximately 80% of the tumor samples. The first malady known to be allied with miRNA degradation was chronic lymphocytic leukemia, and the identified two microRNAs were miRNA-15a and miRNA-16-1 within locus 13q14 (Calin et al. 2002). Since then, many researchers have focused on microRNAs' role in producing genetic disorders through microRNA mediated chromatin reorganization or their implication in different human diseases and their utility as drugs. In Table 1, we listed multiple diseases associated with microRNAs based on the previous works, for instance, upon a variety of inflammatory stimuli, the level of miRNA-155, miRNA-147, miRNA146, and miRNA-9 increases markedly in macrophages. In cancer, two major microRNAs groups are identified: one with putative -suppressive, including the microRNA Let-7 family that plays a crucial role in stopping cancer cells' growth and it is known as a "post-transcriptional-gatekeeper" (Madison et al. 2015). The second group has oncogenic proprieties such as miR-21, which has been discovered as an inhibitor of the tumor suppressor PDCD4, promoting cancer's invasion and metastasis (Asangani et al. 2008). Various microRNAs are reported to be related to neurodegenerative maladies as Parkinson disease (increase level of miR-7 and miR-153) (Leggio et al. 2017) and Alzheimer's disease. For example, a high level of miR-125b stimulated Tau phosphorylation and aggregation by targeting phosphatases PPP1CA and DUSP6, whereas its low level induced a reduction in Tau phosphorylation (Banzhaf-Strathmann et al. 2014). MicroRNAs are also implicated in cell development and proliferation and other diverse pathological processes such as cardiovascular and autoimmune diseases, diabetes, and so more.

This chapter aims to summarize, compare, and analyze various biosensing technologies that have been developed recently towards microRNAs detection and comprehensively highlight recent advances in improving the performance of miRNAs biosensors with signal amplification using functional nanomaterials, nucleic acid circuitry, and/or enzyme.

## 2 MicroRNA Detection

### *2.1 MicroRNA Detection Challenges*

Recently, multiple studies have shed light on microRNAs as biomarkers for the diagnosis of numerous diseases. Hence, the development of reliable and efficient biosensors with -cost strategies for detecting microRNA is required. Due to their unique characteristics, the analysis of microRNAs expression profile exhibits many challenges. Among the latter, the low abundance of microRNAs in real samples since it occupies a tiny fraction compared to the total RNA (around 0.01%) might affect the sensitivity, the duration of the assay and appeal for further enrichment and amplification steps. Add to that their sequence similarity among the same family and their stability to prevent accurate and selective detection. Another challenge associated with microRNAs detection is their small size. This propriety causes difficulties for a selective pairing of miRNAs and improves the cross-hybridization. Finally, the in situ detection may engender a false positive signal if researchers use a mixture of pre-miRNA and miRNA (Cissell et al. 2007). Based on these challenges, there is an ultimate need to develop a sensitive, selective, accurate, and rapid miRNA detection method.

## 2.2 Conventional Techniques Used for miRNA Detection

Northern blotting is the first and most used microRNA detection method and is still considered the gold standard for analyzing microRNA expression profiles. This technique is based on the target miRNA's hybridization captured on a nitrocellulose membrane's surface to a labeled probe. However, the major shortcomings of northern blotting are the high amount of sample loading ($\sim$10–30 $\mu$g), the low sensitivity with a detection limit in the range of nanomolar, and radioactive tags probes (Cissell et al. 2007; Válóczi et al. 2004; Ouyang et al. 2019). To omit these drawbacks, researchers tried to improve the detection of miRNAs via the use of locked nucleic acids (LNA) as a detection probe instead of standard DNA probes. As a result, they found that the sensitivity increases by 10-folds (Válóczi et al. 2004). Moreover, they employed digoxigenin (DIG)—labeling probe as an alternative for the previous radioactive probes. It has since proven that DIG has a short exposure time, high sensitivity, longer shelf life, and increased safety than 32P-labeled probes (Kim et al. 2010; Ramkissoon et al. 2006).

Quantitative PCR preceded by reverse transcriptome of miRNA to cDNA is also classified as a standard strategy for short ncRNAs detection. The major problem associated with this technique is the length of microRNAs that necessitate short primers. The latter may affect the efficiency of the method since it influences the melting temperature (Ouyang et al. 2019). Thus, researchers have made a great effort to solve this problem by amalgamating LNA to the microRNA primer (Pritchard et al. 2015), and combining DNA probe with the ribonucleotides via exploiting miRNA in the role of the template and T4 ligase. With the latter method, Zhang et al. were able to detect as low as 0.2 fM of miRNA (Zhang et al. 2011). They also used digital droplet PCR (ddPCR) since it is more accurate and sensitive in detecting miRNAs (Zhao et al. 2018b). Even though this method has several advantages, it is cumbersome and often used to validate or supplement other techniques (Ouyang et al. 2019). Biochip technology or microarray is another method that was frequently used for microRNA detection. A particular signal is generated to characterize the hybridization event when the labeled DNA probe attached to a solid surface binds to the target miRNA (Ouyang et al. 2019). This technique is designed for rapid and simultaneous detection of different microRNAs. However, it is more expensive, low sensitive, and selective. To ameliorate the detection results, scientists exploit nanomaterials and enzymes as signal amplification techniques. On the other hand, they used LNA instead of DNA probes for Tm normalization and better mismatch discrimination (Castoldi et al. 2006).

## 2.3 Emerging Techniques for microRNA Detection

Over the past decade, substantial scientific investments have been made to improve conventional methods and build up a suitable selective, sensitive, and high-throughput platform to detect miRNAs such as biosensors. Recently biosensing gains great interest for the detection of microRNAs. Biosensor is a bioanalytical device that permits converting a biological reaction into a measurable signal in which there are different forms of transducers, including optical, electrochemical, and electrical (field-effect transistor and nanopore sensing). In this part, we will deal with these different strategies for the detection of microRNA.

### 2.3.1 Optical Biosensing

Optical assays have been considered an emergent tool for miRNA detection due to multiple advantages, including high sensitivity, convenience, and uncomplicatedness. Four types of assays are mainly involved in the optical sensing, fluorescent, colorimetric, surface-enhanced Raman scattering (SERS), and surface plasmon resonance (SPR). This section will include different optical biosensor for micoRNAs detection.

(1) Different probes used in optical sensing for microRNAs detection

Linear single-strand DNA/LNA/PNA represents the most used biological recognition unit for microRNA detection (Dave et al. 2019). The LNA probe possesses a high affinity towards small and highly similar targets and high thermal stability compared to traditional DNA probes. Therefore, it allows for better discriminatory function and better sensitivity (Silahtaroglu et al. 2004). PNA probe also gained great interest by scientists due to its advantages, including its faster hybridization, stability, and resistance to enzyme degradation (Vilaivan 2018). Molecular beacons (MBs) is another different form of microRNA probe characterized by a hairpin shape. It is also widely used for microRNAs detection. This probe is generally labeled from both ends, one modified with a fluorophore and the second end tagged with a quencher. Once this probe is hybridized with the target microRNA, the short double-strand will be dissociated to generate a linear form, thus restoring its fluorescence (Tyagi and Kramer 1996; Mittal et al. 2019). MB is an excellent alternative for multiplexed microRNAs detection (Lee et al. 2016) though; its major limitation is associated with the depleted quenching effectiveness (Shu Zhu et al. 2019). A Y-shaped DNA probe can also be investigated for microRNAs detection. It is formed when three oligonucleotides; two probes, and a target DNA complementary binds to each other. This kind of probe is sensitive to single base mismatch (Zhang et al. 2018; He et al. 2016).

On the other hand, one of the most used DNA nanostructures is a tetrahedron with six double-helical edges, assembled by annealing four DNA strands with partial complementarity. DNA tetrahedron have been used widely in microRNA detection in recent years (Xu et al. 2016). This DNA shape has various characteristics, such as good rigidity, enzyme resistance, and excellent cellular permeability (Goodman et al. 2005). The choice of the right probe is based on the application itself and the craved results.

(2) Colorimetric sensor

Attributable to its advantages, including low cost and quick response, colorimetric biosensor has become an appealing microRNA detection technique. It is based on organic chromogenic substrates, such as ABTS and TMB to form colored products via enzymatic catalysis (Zhu and Gao 2018). However, this type of biosensor lacks sensitivity and selectivity compared to others. Hence, noble metal-based localized surface plasmon resonance (LSPR) represents a better alternative for the traditional colorimetric sensor for microRNAs detection since its first use by Joshi et al. (2014), Bellassai et al. (2019). In most cases, researchers profited from AuNPs for the LSPR based colorimetric sensor since they can control their colloidal aggregation and dispersion phases by utilizing the target analytes. The color of the solution and the aggregation of nanoparticles are mainly influenced by the balance between particle attraction and repulsion forces. In the case of aggregation, the reason can be attributed to the "interparticle-crosslinking" or the "no crosslinking-aggregation" mechanism. In this context, Huang et al. designed a colorimetric experiment for the detection of miRNA-21 using AuNPs and DSN enzyme for better signal amplification. After the different amplification cycles, the aggregation of AuNPs is derived by the use of divalent cations ($Mg_2^+$). The detection limit was about 50 pM with a linear range from m 50 pM to 1 nM. This method shows great discrimination results even for a single mismatch and it reveals high performance for biomedical application since it was tested for cell lysates (Fig. 4) (Huang et al. 2019).

(3) Surface plasmon resonance (SPR) sensor

Surface plasmon resonance (SPR) represents the fact where the electrons of the surface metal layer are excited via the light photons with a certain incidence angle and then spread parallel to the metal layer (Fig. 2c). A slight variation in the reflective index will occur in the analyte presence, making it possible for microRNAs detection (Zeng et al. 2017). Nanomaterials represent an ultimate strategy for signal amplification in SPR based biosensors. Mouzavi and his co-workers succeeded in developing a dual hybridization biosensor based on the SPR on capped gold nanoslits using magnetic nanoparticles as a target enrichment tool and signal amplification strategy to detect the urinary miRNA-16-5p. They were able to reach 17 fM as a limit of detection. This technique allocates the detection of the target miRNA in the urine of acute kidney injury patients (Mousavi et al. 2015). Li and his co-workers developed an SPR biosensor composed of two layers of graphene oxide-gold nanoparticles (GO-AuNPs) composites (Fig. 2b). The first layer of the nanocomposite functionalized with the capture probe was immobilized onto the surface of Au film modified with thiocyanuric acid In contrast, the upper layer (used for amplification) is modified with an assistant probe. In the presence of the target miR-141, both layers are joined (Fig. 2a). The detection limit reached 0.1 fM, and the selectivity was tested in the presence of different miRNA-200 family members (Li et al. 2017). Another study by Wang's group wherein they resorted to an SPR platform employing gold film on which is immobilized the capture probe for miRNA-141 detection and gold nanoparticles-decorated molybdenum sulfide where the assisted DNA that binds to

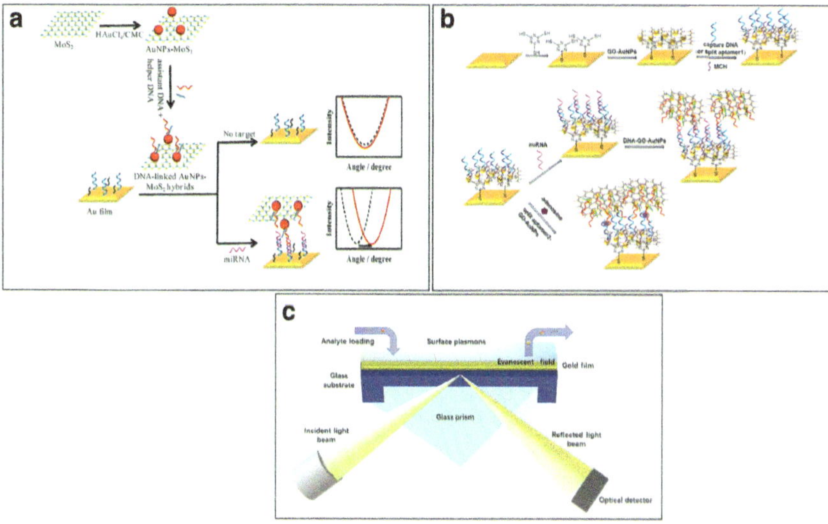

**Fig. 2** **a** SPR biosensor for microRNA-141 detection using AuNPs-MOS2 hybrids for signal enhancement **b** SPR biosensor for microRNA-141 detection using double layers of graphene oxide-AuNPs for signal enhancement **c** SPR detection principle. Reproduced with permission (Nie et al. 2017; Li et al. 2017)

a segment of miRNA-141 is linked. This enzyme-free method displays high sensitivity with a limit of detection equal to 0.5 fM and high selectivity. Moreover, this method has an excellent capacity for complex samples, such as human serum (Nie et al. 2017).

(4) Surface-enhanced Raman scattering

SERS is a surface-sensitive method based on Raman scattering enhancement via the use of irregular metal shells or nanostructures. Considering that the signal boost may attain 1011-fold, SERS has become a captivating tool for disease-related microRNAs detection. For SERS, two significant theories exist; the electromagnetic theory relies on the excitation of localized surface plasmons and the chemical concept that aims to create charge-transfer complexes. Guo et al. designed a sandwich-type method for simultaneous detection of three different microRNAs (miR-21, Let-7d, and miR-141). In this study, they fastened thiol-containing Raman dyes on Au@Ag nanosnowmen nanoprobes. They employed the gold substrate in SERS-active substrates' role, wherein they fixed the different capture probes. The limit of detection was 1.089 fM, 0.839 fM, and 1.019 fM, respectively (Guo et al. 2018). Some groups took advantage of combining the use of nanomaterials with the enzyme-based signal amplification such as DSN and exonuclease to improve the sensitivity. For instance, Yang and his colleagues developed a sensitive platform for miR-155 detection based on TB@$CaCO_3$ loading DNA microcapsule with the SERS active substrate Si@Nafion@Ag. In the presence of the target miRNA and via the aid

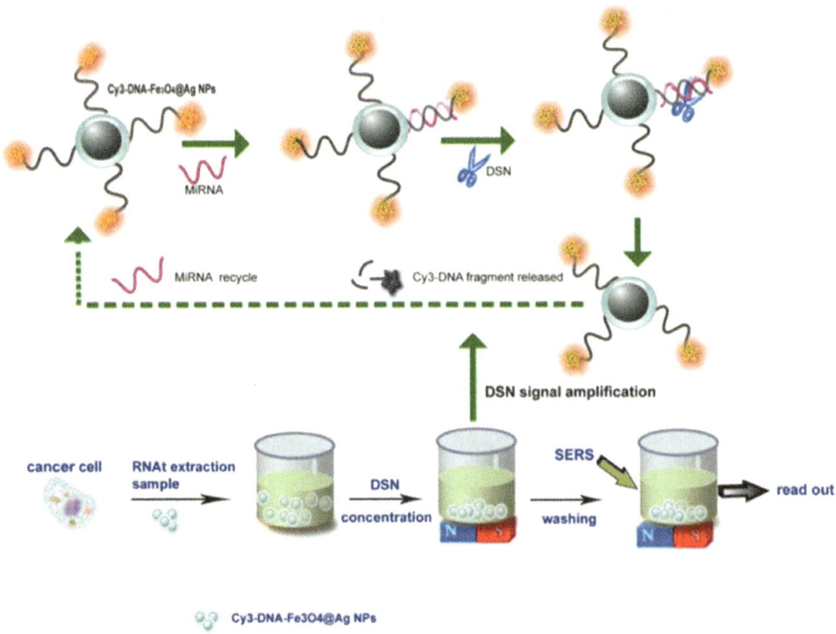


**Fig. 3** SERS detection of the miRNA Let-7b using $Fe_3O_4$@Ag nanocomposites and DSN enzyme for signal amplification. Reproduced with permission (Pang et al. 2016)

of enzymatic amplification strategy using duplex-specific nuclease, the Raman dye (TB) was discharged from TB@$CaCO_3$ composite. The achieved detection limit was about 0.67 fM (Yang et al. 2018). Zhu's group created a "signal-off" platform for miRNA-21 detection by incorporating the T7-Exo enzyme as a recycling amplification approach along with gold decorated silicon nanowire arrays. To ensure further enhance the sensitivity of their SERS sensor; they employed Au nanostar probes to provide a stronger electromagnetic field. Via this technique, they were able to detect as low as 0.34 fM (Wen et al. 2019). Another study conducted by Xiao's group for Let-7b detection relies on the immobilization of DNA probe onto the surface of $Fe_3O_4$@Ag NPs. When the hybridization with the target miRNA takes place, the DSN initiates the target recycling cycles. The nanoparticles were collected using a permanent magnet, and the SERS signal was measured (Fig. 3). This technique possesses a fabulous limit of detection (0.3 fM) and an excellent recovery while checking it for different cell lines (Pang et al. 2016).

(5) Fluorescence-based sensor

Fluorescence assay is the most predominantly method used for microRNA biosensing (Ye et al. 2019). It depicts the fact that a fluorescent dye or molecule absorbs and emits light. In the absorption phase, the fluorescent tag is excited by light with high-energy (short wavelength), stimulating electrons' transition within the fundamental

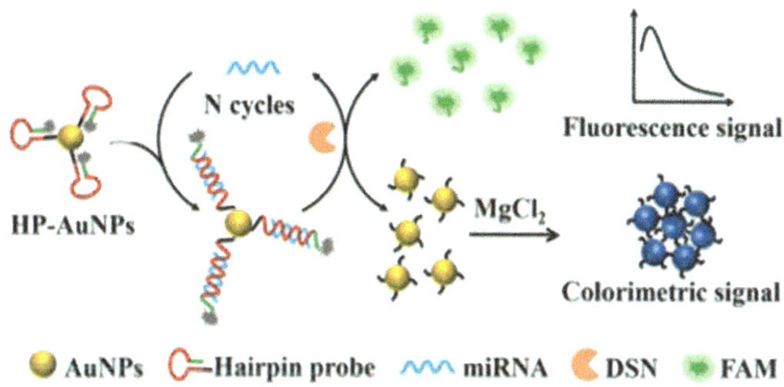


**Fig. 4** Principle of the fluorescent and colorimetric biosensor using AuNPs designed for microRNA-21 detection based on DSN assisted target recycling. Reproduced with permission (Huang et al. 2019)

state's molecule to the excited state. Once in this state, and within nanoseconds (the fluorescence lifetime), the electrons will release their stored energy and rest back to the ground state, generating an emitted photon with lower energy (longer wavelength) compared to the absorbed light (Zhu and Gao 2018). Recently, a new fluorescence approach called fluorescence resonance energy transfer or Förster resonance energy transfer (FRET) has gained great interest in miRNAs detection. This strategy relies on energy transmission from a fluorescent donor to an acceptor. Two main conditions are needed for FRET: (i) the donor's emission wavelength corresponds to the fluorescent acceptor's excitation wavelength, and (ii) the distance between them does not exceed 10 nm (Sapsford et al. 2006).

Attributable to the effect of the considerable limitations, many nanomaterials can directly engender fluorescent emissions with the exalted significant harvest, excellent photo-stability, and long fluorescence lifetime, including quantum dots (QDs). Jie et al. reported a multiplexed fluorescent sensor for detecting miRNA-141 and miRNA-21 using two different kinds of quantum dot (CdSe@ZnS and CdTe) probes and target recycling amplification strategy. The thiolated hairpin DNA probe was immobilized onto the magnetic bead's surface decorated with Au nanoparticles (MB@Au). During hybridization with the target miRNAs, Exo III-based target recycling is unthreaded, generating short ssDNA sequences attached to the surface of MB@Au. Afterward, the latter is hybridized with QDs fluorescent probes that will activate the cleavage by adding Nb.BbvCI enzyme followed by magnetic separation and underwent for fluorescence measurements. The achieved detection limit for miR-21 and miR-141 was 1.5 pM (Jie et al. 2017).

Nanomaterials can also serve as effective fluorescence quenchers (Fig. 5). For instance, some of these nanomaterials, such as gold nanoparticles (AuNPs), (Huang et al. 2019) graphene, (Tan et al. 2019) $WS_2$, (Xi et al. 2014) were investigated to detect miR-21 using DSN based signal amplification. The calculated limit of detection was 50 pM, 1.5 pM, and 300 fM, respectively. Moreover, the same approach for miRNA Let-7a detection was used employing metal nanomaterials such as a

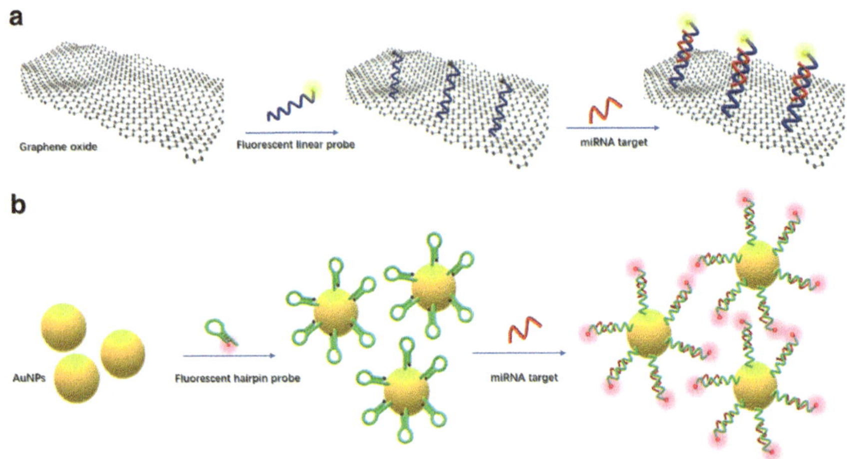


**Fig. 5** Principle of the fluorescence quenching: **a** using graphene oxide, **b** using gold nanoparticles

magnetic bead, and they were able to detect as low as 300 fM (Shen et al. 2015). Xie group reported a fluorescent-based biosensor to detect miRNA-141 using graphene oxide and rolling circle amplification along with Exo III. This method exhibits a good selectivity and sensitivity along with high performance when testing the spiked miRNA in the human serum, and the LOD was about 0.1 aM (Li et al. 2019). All these methods show tremendous results for the selectivity test towards mismatches and different miRNAs families. The exploitation of nanomaterials for fluorescent sensors not only expands the selectivity, the sensitivity, and the performance of the detection of clinical biomarkers (miRNA), but also it opens the door to plenty of novel approaches to conceive fluorescent sensors.

### 2.3.2 Electrochemical Biosensor for miRNA Detection

Electrochemical methods exhibit superior benefits own to their capability of miniaturization, selectivity, and capability compared to other analytical techniques, which uplift it to have widespread use in agriculture, food, environment fields, and medical application (Bahri et al. 2019). Also, electrochemical biosensors demonstrated pleasant for serving simple, fast, and point of care analysis. The transduction element for such a genosensor can be in multiple materials: glassy carbon, gold, indium tin oxide, graphite, or the modification of these materials using multiples nanoparticle, nanotubes and/or nanowires. The fundamental principle in the electrochemical detection of miRNAs is measuring the variations either in the electrode properties (resistance or capacitance) or in the electrochemical tag's signal against the target miRNA hybridization with the complementary probe.

(1) Impedimetric miRNA biosensors

Electrochemical impedance spectroscopy (EIS) is a useful detection technique that delivers crucial information while modification processes like; hybridization, nanostructure formation, and bio-functionalization. Furthermore, EIS is a powerful tool for surface adsorption explore and electrochemical reaction mechanisms investigation (Tlili et al. 2005, 2003) . When it comes to miRNA detection, EIS has been employed because it is label-free and practical. Some works detail EIS as a detection strategy for miRNA (Erdem et al. 2017) while others describe it as a characterization method (Erdem et al. 2015). Peng and Gao (2011) suggested an electrochemical impedance-based biosensor using a novel sensing protocol via the amplification of miRNA's electrochemical detection and greatly enhanced sensitivity and specificity. The developed biosensor was based on a direct ligation procedure through a direct tag of miRNAs with ruthenium oxide nanoparticles ($RuO_2$ NPs). The $RuO_2$ NPs-initiated polymerization of 3, 3'-dimethoxybenzidine (DB), and the hybridized miRNA strands and free capture probe (CP) strands guided the deposition of poly(3, 3-dimethoxybenzidine) (PDB). Whereas the biosensor surface was made of a mixed monolayer of 4-mercaptoaniline and oligonucleotide capture probes on a gold electrode. Following this protocol, and after 60 min of incubation into $DB/H_2O_2$ mixture, the developed biosensor showed a linear charge transfer resistance-concentration relationship from 6 fM to 2 pM and a detection limit (LOD) of 3 fM (Fig. 6a).

Similarly, using PDB polymerization strategy, Gao and co-workers suggested another label-free detection of circulating miRNAs in blood along with miRNAs extracted from cultured cells with a detection limit of 2 fM (Gao et al. 2013a) (Fig. 6b). Graphene oxide modified pencil graphite electrode (PGE) was also utilized for electrochemical detection of miRNA-34a and showed appealing discrimination of the target miRNA against miRNA-155, miRNA 660, and miRNA-15a in buffer medium and in the fetal bovine serum (Congur et al. 2015) (Fig. 6c). In another work, the graphene-modified disposable pencil graphite electrode (GME) was utilized by Ozsoz team (Kilic et al. 2015), wherein they used voltammetric and impedimetric detection techniques for the detection of microRNA-21. They reported a high performance of the suggested technique toward miRNA-21 in both buffer and cell lysates samples with a LOD of 2.09 μg/mL. Zhang et al. (2016) developed an electrochemical impedance biosensor with immobilization-free for amplified detection of miRNA-21. In their work, they have used a strategy based on a duplex-specific nuclease (DSN) assisted target recycling and capture probe enriched from the solvent to the magnetic glass carbon electrode surface (as a working electrode) using magnetic beads. Due to the low activity of duplex-specific nuclease (DSN) to ssDNA, capture probes cannot be hydrolyzed in the absence of miRNA-21. However, in the presence of the target miRNA-21 forming DNA-RNA heteroduplex through the hybridization with the capture probes, the DSN hydrolyzed the target-binding portion of the capture probe while liberating the intact miRNA-21 to be hybridized with another capture probe and a second hydrolization cycle is starting. Finally, all capture probes were digested. Moreover, real sample measurements were made using human serum samples from breast cancer patients selective attempts toward miRNA-21 detection. By

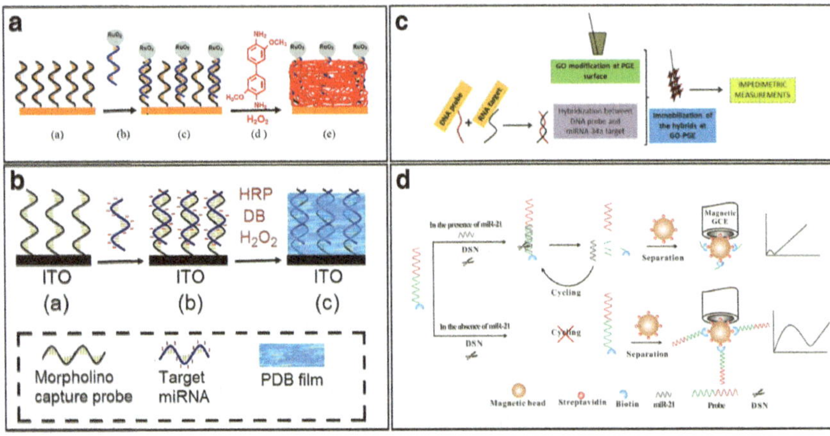


**Fig. 6** Examples of impedimetric miRNA biosensors strategies. **a** Experimental principle of the amplified detection of miRNA based on $RuO_2$ NP. **b** Schematic illustration of the label-free biosensor for electrochemical detection of femtomolar microRNAs. **c** The working strategy of the impedimetric Detection of microRNA at Graphene Oxide Modified Sensors. **d** Experimental principle of immobilization-free electrochemical impedance biosensor based on DSN assisted target recycling for amplified microRNA detection. **a** Reproduced with permission (Peng and Gao 2011), **b** Reproduced with permission (Gao et al. 2013a), **c** Reproduced with permission (Congur et al. 2015), **d** Reproduced with permission (Zhang et al. 2016)

employing such a strategy the developed biosensor displayed an ultrahigh sensitivity for miRNA-21 with a LOD of 60 aM (Fig. 6d).

(2) Amperometric miRNA biosensors

Amperometry is an electroanalytical method based on the current resulting from the application of a constant oxidizing or reduction potential and respect to time to a working electrode (indicator). Usually, the magnitude of the obtained current depends on the concentration of the oxidized or reduced substance. Thus this method can be used for various analytical applications.

*(a) Chronoamperometry*

Chronoamperometry is a time-dependent method where the square-wave potential is applied to the working electrode. The current to time measurement of the electrode fluctuates depending on the diffusion of the solution immersed in the bulk analyte toward the sensor surface. Therefore, chronoamperometry can be applied to detect current-time dependence for the diffusion-controlled process occurring at an electrode, which varies with analyte concentration. For example, Liu et al. (2014) developed a label-free and highly-sensitive strategy using triple signal amplification via AuNps, AP and p-aminophenol (p-AP) redox cycling for miRNA detection into a range of 10 fM–5 pM (Fig. 7a), with a limit of detection (LOD) of 3 fM. The followed strategy was based on the difference between RNA and DNA structures. The DNA probes were first immobilized onto the Au electrode, then it hybridized with

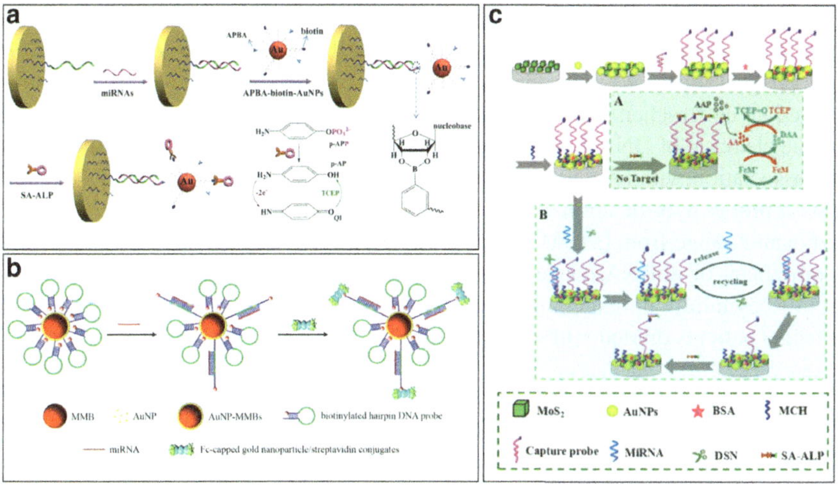


**Fig. 7** Examples of electrochemical detection of miRNA strategies. **a** Schematic illustration of the label-free detection of miRNA using triple signal amplification of APBA-biotin-AuNPs, SA-ALP, and the p-AP redox-cycling reaction. **b** Amplified voltammetric detection of miRNA via the emergence of conducting magnetic microbeads and ferrocene-capped gold nanoparticle/streptavidin conjugates. **c** miRNA-21 detection using signal amplification of DSN and ECC redox cycling via AuNPs/$MOS_2$ modified electrode. **a** reproduced with permission (Liu et al. 2014), **b** reproduced with permission (Lu et al. 2016), **c** reproduced with permission (Shuai et al. 2017)

the target miRNA. The cis-diol ribose sugar group at the end of the miRNAs chain interacted with the 3-aminophenylboronic acid (APBA)/biotin-modified AuNPs via a boronate ester covalent bond. The obtained complex was then consented to react with the streptavidin-conjugated alkaline phosphatase through the biotin-streptavidin interaction. After accumulating the 4-aminophenylphosphate (p-APP) substrate, the enzymatic conversion from p-APP to p-AP developed. The resulting p-AP possibly cycled via a chemical reducing reagent after its electro-oxidization on the electrode, thus increasing the anodic current. In another work, Castaneda et al. (2017) combined the electrocatalytic amplification (ECA), and duplex-specific nuclease (DSN) catalyzed amplification for the detection of miRNA. In their work, they have used Pt nanoparticles to enhance the catalytic electrochemical reaction on inert Au ultramicroelectrodes (UMEs) after miRNA-capture probe hybridization followed by DSN enzyme reaction on DNA.

*(b) Chronocoulometry*

Compared to chronoamperometry, Chronocoulometry is also similar, except that the readout is a charge variation in function of time. Chronocoulometry exhibits the advantages of effective integration in terms of reducing the noise signal, where it is easy to distinguish and separate the capacitive charge and the faradic charge. Masud et al. (2017) developed a nonenzymatic, amplification-free, and highly sensitive miRNA based on gold-loaded nanoporous iron-oxide nanocubes

($Au@NPFe_2O_3NC$). The assay displayed a remarkable selectivity toward miRNA21 with a detection limit down to 100 fM in cell lines and tissue samples from oesophageal squamous-cell carcinoma patients. Whereas, Miao et al. (2016) demonstrated an electrochemical miRNA biosensor based on DSN enzyme, which relied on the cleavage of DNA-RNA hybrid duplex cleavage by DSN and the miRNA target association with $[Ru(NH_3)_6]^{3+}$ upon introduction of AuNPs due to DNA1 and DNA2 probes hybridization. The used method exhibited a LOD as low as 50 aM into a dynamic range from 0.1 fM to 100 pM. This direct detection method without the requirement of miRNA conversion to cDNA also showed excellent discrimination of miRNA family in real complex sample formed of throat swabs H1N1 influenza-infected patients diluted with sterile saline.

(3) Voltammetric miRNA biosensors

Voltammetry method is classified as an electroanalytical technique, where the current measurement is achieved at a potential ramp. The wide variety of methods in which the applied potential may be varied leads to numerous type of voltammetric techniques, including Cyclic voltammetry (CV), Square wave voltammetry (SWV), and Difference pulse voltammetry (DPV) (Rezaei and Irannejad 2019).

*(a) Cyclic voltammetry (CV)*

CV is widely used to get useful information about redox potentials and explore the mechanisms and kinetics parameters involved in electroactive analyte solutions' reactions. When it comes to miRNA detection CV technique can be used for label-free and labeled detection. Label-free detection of miRNAs is generally a simple method where no additional labeling steps are required, making it less time consumption, more practical, and usually low cost compared to label-based methods. The plurality of the label-free miRNA biosensors approaches is based on identifying electroactive nucleic acid base signals before and after hybridization. AuNPs have been widely used and so popular for nucleic acid-based detection strategies. Following the same strategy of dsDNA and ssDNA's adsorption ability on AuNPs, Li et al. (2016) developed a let-7a miRNA biosensor with a LOD of 16 fM in human breast adenocarcinoma cells. Lu et al. (2016) reached a lower LOD of 0.14 fM using AuNPs based method and the higher loading density of biotinylated hairpin-structured DNA probes that opened after hybridization and further interacted with ferrocene-capped streptavidin-conjugated (Fig. 7b).

*(b) Square wave voltammetry (SWV)*

SWV technique is one of the most sensitive and fastest pulse voltammetry techniques. The obtained detection limits are comparable with those obtained using spectroscopic and chromatographic techniques (Simões and Xavier 2017). In one study, Tran et al. (2013) have used the SWV to detect miRNA-141 by taking the advantage of the quinone group existing inside the nanostructured polymer film containing an electroactive polymer and carbon nanotube. The presence of the target microRNA-141 generated a signal-on response resulting from the enhancement of the polymer. The detection of prostate cancer biomarker mir-141 was achieved using square wave voltammograms leading to a LOD of 8 fM. In a similar context, Labib et al. (2013) designed a three-mode experience electrochemical biosensor based on hybridization, protein displacement, and P19 protein binding for the detection and quantification of miRNAs with a LOD of 5 aM without the requirement of PCR amplification. MiRNA biosensors-based hybrid nanomaterials also showed a greatly enhancement of sensitivity and limit of detection. For example, an interesting hierarchical flower-like Au nanostructure that exhibits a large surface area and thus enhances the sensitivity up to 1 fM also enables discrimination of the target miRNA from cervical cancer cells (HeLa) and lung cancer cells (A549) (Su et al. 2016). Besides the AuNPs, ruthenium nanoparticle ($RuO_2$) was also used for miRNA detection along with high conductivity and a robust catalytic activity (Peng et al. 2010).

*(c) Difference pulse voltammetry (DPV)*

DPV is a technique that implicates applying amplitude potential pulses on a linear ramp potential. Using DPV, a base potential value is selected at which there is no faradaic reaction and is applied to the electrode. The bias potential is raised between pulses with equal increments. The current is instantly measured before the pulse application and at the end of the pulse. Consequently, the difference between them is registered. Label-free detection of miRNA based on DPV measurements was also widely used, leading to high discrimination of the target biomolecules along with high sensitivity and a low limit of detection (LOD). Kilic et al. (2013) suggested an alternative procedure in which they used the P19 viral protein as a bio-recognition element for label-free electrochemical detection of miRNA-21 due to its particular binding ability RNA duplexes. The developed mir-21 biosensor was based on the oxidation signal of tryptophan in p19 protein afore and after interaction of the protein with miRNA hybrid and displayed a picomolar detection limit in real samples measurement. The selectivity test was proven with two control experiments, firstly via a non-complementary miRNA (miRNA-192), which leads to a non-hybridization or p19 sequencing. Whereas the second test was based on glucose oxidase enzyme (GOX) instead of p19, confirming the specificity of protein toward the double-stranded RNA (dsRNA).

Yang et al. (2009) also developed a label-free electrochemically inactive inosine-substituted probe sequence that was utilized to get an assay with a yes/no signal when the target miRNA was detected using the DPV technique. Isin et al. (2017) modified the graphene oxide onto the surface of a pencil graphite electrode (PGEs) and the modified electrode was used for the first time for voltammetric monitoring of miRNAs-34a biomarker related to Alzheimer disease. The CA/GO/PGE characterization was investigated via CV, EIS, and scanning electron microscopy (SEM). The GO concentration, DNA probe concentration, and miRNA-34a were optimized. Thus, under the optimum conditions, the developed biosensor exhibited a LOD of 7.52 μg/mL into a linear concentration range from 5 to 35 μg/mL.

Although these methods display a more straightforward, fast solution for miRNA detection and quantification, but still required sensitivity enhancement that could not be met most of the time, modifying the surface properties or the electrode's conductivity via metal nanoparticles, electroactive materials or taking advantage of enzymes for the signal amplification could significantly lower the limit of detection and enhance the biosensor's sensitivity. Electroactive materials such as methylene blue (MB) and meldola's blue (MDB) have been employed as electroactive redox labels for hybridization indicators in electrochemical nucleic acid as well as miRNA biosensors. Using DPV technique, Kilic et al. (2012) studied miRNAs detection based on alkaline phosphatase (AP) enzyme. In this work, kilic and coworkers described an assay usage of (PGE) for oligonucleotide capture probe immobilization via EDC/NHS chemistry to record the $\alpha$-naphthol oxidation signal of AP enzyme after the hybridization with target mir-21. Similarly, Rafiee et al. (2016) developed a simple electrochemical miRNA biosensor using methylene blue (MB) as an electroactive intercalator.

The DPV technique was used to study the oxidation of MB associated with the hybridization event between the probe DNA and the target miRNA-21. This approach displays a high selectivity and sensitivity with a LOD of about 84.3 fM. In another study, Kapton et al. (2017) reported an electrochemical sensor for the detection of mir-21 extracted from breast cancer cell (MCF-7) based on Meldola's blue (MDB) reduction. Wherein, they employed the electropolymerized polypyrrole modified pencil graphite as an electrode (PPy/PGE). This method displayed a high sensitivity and a LOD of 0.17 nM.

Hybrid nanomaterials also showed superior features for miRNA detection, like molybdenum disulfide ($MOS_2$) microcubes (Shuai et al. 2017) and tungsten oxide-graphene composites (Shuai et al. 2016). Shai et al. (2017) designed an ultrasensitive electrochemical biosensor for miRNA detection using hollow molybdenum disulfide ($MOS_2$) microcubes and DSN enzyme was used for signal amplification.

Firstly, the biotinylated ssDNA probe was immobilized onto AuNPS/$MOS_2$ modified electrode for streptavidin-conjugated alkaline phosphatase combination (Fig. 7c). When the DSN cleaves the formative duplexes of the capture probes and miRNAs, the biotin group strips from the electrode's surface, and the streptavidin-conjugated alkaline phosphatase is not able to be attached to the electrode surface. Subsequently, ascorbic acids create the electrochemical-chemical-chemical redox cycling and making the electrochemical response in the attendance of ferrocene methanol and tris (2-carboxyethyl) phosphine. The suggested method exhibits a LOD of 0.086 fM into a concentration range from 0.1 fM to 0.1 pM. Additionally, the biosensor showed a successful capability to detect target miRNA-21 in human serum samples.

### 2.3.3 Field-Effect Transistor (FET)

Additionally to the electrochemical detection method, field-effect transistor (FET) is also a prominent part of the electrical detection. Recently, label-free detection based FET biosensors has shown great attention and prospects considering no electrochemical tags. However, attain high capability detection and sensitivity. Gao et al. (2013b) reported the design of an ultrasensitive, real-time, and label-free detection of miR-21 using silicon nanowire (SiNWs) field-effect transistor biosensor array. The sensing channel of silicon nanowires was synthesized via optical lithography and anisotropic self-stop etching. Further, the obtained device was functionalized with 3-aminopropyltriethoxysilane (APTES). The terminal carboxyl group of the capture probe DNA was then conjugated to the amine group of the APTES-modified SiNWs via EDC/NHS chemistry. The described strategy leads to the detection of miR-21 with a LOD of 1 fM and high selectivity for single-nucleotide polymorphism discrimination.

Graphene a 2D sheet of $sp^2$ bonded carbon atoms sparked research on 2D materials that are blooming at a tremendous rate owing to its unique properties, such as high carrier mobility, low electrical noises, ease of fabrication and functionalization, and large active detection area (Hwang et al. 2020). Song et al. (2020) developed a 3D graphene FET (GFET) biosensor, ssDNA as a probe was immobilized with 1-pyrenebutanoic acid succinimidyl ester (PBASE) as a molecule linker.

The manufactured biosensor showed a linear detection response in a concentration range of 100 pM–100 nM with high sensitivity of about 35 mV/nM and a LOD of 100 pM. Femtomolar miRNA detection was also proposed by Zhang team (Cai et al. 2015) based on GFET decorated gold nanoparticles.

The emerge of gold nanoparticles with peptide nucleic acid (PNA) employment instead of DNA leads to a higher hybridization efficiency with a LOD as low as 10 fM. What's more, the obtained assay exhibits accurate discrimination of complementary miRNA from non-complementary miRNA and one-base mismatched miRNA. Very recently, Gao et al. (2020) reported a free labeling and flexible GFET biosensor along with robust performance, specific and ultrasensitive detection of miRNA. The DNA probes were immobilized onto the graphene channel via $\pi - \pi$ stacking interaction without molecules linker or surface functionalization. The proposed biosensor finished the miRNA detection in only 20 min and displayed a detection ability down to 10 fM.

Molybdenum disulfide ($MOS_2$) as a transition metal dichalcogenides (TMDCs) material has also added exceptional value to the electrical measurement-based biosensors field-effect transistor. $MOS_2$, with its direct-to indirect tuning structure as varying its layer number from single layer to bulk, respectively (Sarkar et al. 2014). Numerous studies on $MOS_2$ have fully emphasized its extraordinary potential for electronic components. Also $MOS_2$-based biosensors for DNA or proteins detection have been reported (Sarkar et al. 2014). Furthermore, a label-free ultrasensitive biosensor platform using $MOS_2$ FET gadgets were developed for breast cancer biomarker (miRNA-155) in human serum and cell line samples (Majd et al. 2018). The developed biosensors displayed high mobility of $1.98 \times 103\ cm^2V^{-1}\ s^{-1}$ and a LOD of 0.03 fM under a complementary target miRNA-155 concentration range from 0.1 fM to 10 nM. Table 2 is resuming and illustrating the conventional development methods of miRNA biosensors based field-effect transistors.

**Table 2** Conventional development methods of miRNA biosensors based field-effect transistors

| FET structure | Sensing material | Surface functionalization with | DNA probe | miRNA | Work example | LOD |
|---|---|---|---|---|---|---|
|  SiNW FET | 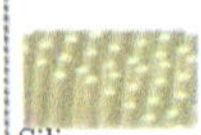 Silicon Nanowires | APTES + $NH_2$ Probe |  | 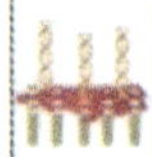 | Gao et al., 2013 | 1 fM |
|  GFET |  Graphene | AuNPs + SH Probe |  |  | Cai et al., 2015 | 10 fM |
| | | PBASE + $NH_2$ Probe |  |  | Song et al., 2020 | 100 fM |
| | | Adsorption |  | 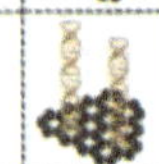 | Gao et al., 2020 | 10 fM |
|  $MoS_2$ FET |  TMDCs ($MoS_2$) | Adsorption |  |  | Majd et al., 2018 | 0.03 fM |

### 2.3.4 Nanopore Technology

Nanopore technology has a strong potential to enhance the development of low coast high-performance single-molecule sensing platforms. Nanopore-based sensors have been successfully used as a detection platform for different biological targets, such as DNAs, aptamers, and MicroRNAs (Wanunu 2012; Zhao et al. 2018a; Sultan and Kanavarioti 2019). The detection principle of nanopore-based sensors is inspired from the Coulter classical counter routine; a nanopore chip is embedded inside a flow cell between two chambers (cis and trans). The two reservoirs are filled with an ionic buffer (KCL, LiCl, $CsCl_2 \cdots$). The application of an electric potential through 2 Ag/AgCl electrodes will force electrons to move from one chamber to another, establishing a current baseline called open pore current. When the target molecule is added to one side of the flow cell, the molecules will be driven to pass through the pore under the applied potential. The passage of the molecules through the pore will temporarily block the pore and prevent electrons from circulating. However, a current drop will appear in the as-established baseline current whenever a molecule passes through the pore. The main parameters to extract from a nanopore sensor are the current drop intensity and the dwell time, which are the characteristics of the target (Lee et al. 2018). Nanopore sensors are divided into two families biological and synthetical nanopores (Fig. 8). The first family is based on a protein channel embedded directly within the lipid bilayer. In contrast, the second one is founded on a nanometric pore drilled in a free-standing membrane (Shi et al. 2017).

(1) Biological nanopore

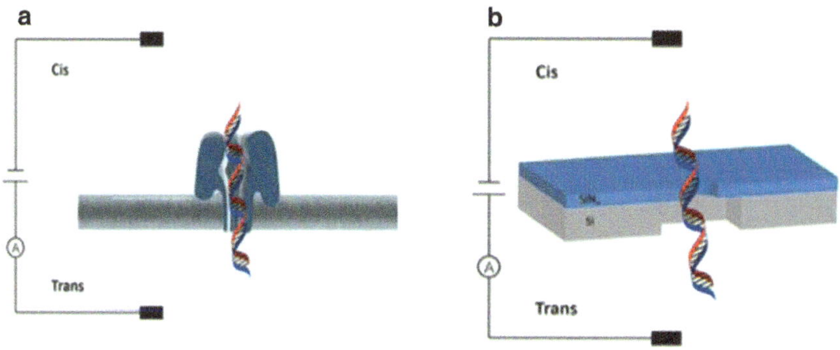


**Fig. 8** **a** Biological nanopore. **b** Solid state nanopore

So far, several studies have been published based on biological nanopores to detect different microRNAs. In one study Ryuji Kawano and his team developed an alpha-hemolysin ($\alpha$HL) nanopore-based miRNAs sensor that utilizes isothermal amplification and asymmetric nanopore measurement for indirect and label-free detection of miRNA-20. In their work, they have reported the detection of miR-20a with a limit of detection (LoD) of 1 fM. This ultra-low concentration was mainly achieved by designing the three-way junction structure with the catalytic enzyme reaction using miRNA-20 as the input and poly-thymines (polyT20) as outputs molecules. This approach can be applied for ultra-low concentration detection of other microRNAs biomarkers by simply changing the nucleotide sequences of the DNA template and primer (Zhang et al. 2017). Another work reported by Ivica et al. was based on $\alpha$-hemolysin pore for the detection of miR155, which is considered as a lung cancer biomarker. In their work, the target MicroRNA was firstly hybridized with a specific probe. Then, the prob-miRNA duplex was unzipped before translocating through the pore. In order to improve the analytical performance of their nanopore assay, they have studied the effect of different salt gradients between the nanopore chambers and the DNA probe design effect. Based on their funding, the 8-fold KCl gradient enabled a linear relationship between pulse frequency and miRNA concentration in the range of 100 pM–100 nM with a limit of quantification of 100 pM (Ivica et al. 2017).

For the sake of selective simultaneous detection of different microRNA targets by using a biological nanopore sensor, Li-Qun Gu and his team have done innovative work and succeed in detecting four different microRNAs miR-155, miR-182-5p, miR-210, and miR21. The approach is based on designing different barcodes probes and attach each one to a target microRNA.

However, the interaction between the probe and the target could modulate the ion flow through the nanopore. Resulting in a specific signature attributed to each complex (barcode probe + target) (Zhang et al. 2014). Moreover, a novel method based on an $\alpha$-hemolysin pore was proposed by Tian et, al. to selectively detect

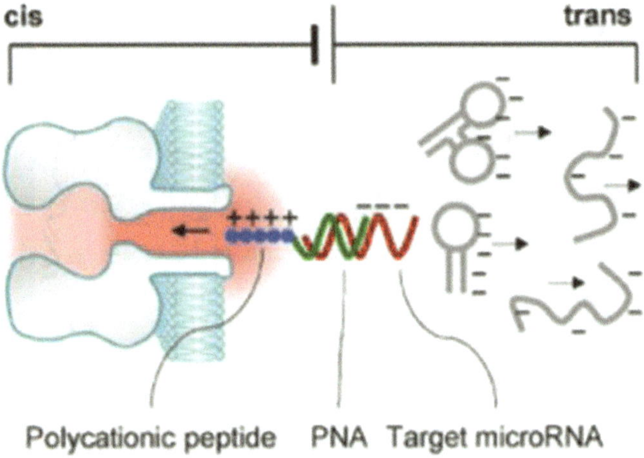


**Fig. 9** Cationic peptide-PNA probe hybridized microRNA selective detection principle. Adapted with permission from (Tian et al. 2013)

microRNA in a nucleic acid mixture (Fig. 9). They have proposed to form a dipole complex by hybridizing a polycationic peptide-PNA probe with the target microRNA.

However, the application of a voltage bias, opposed to the bias needed to capture the nucleic acids, will drive the complex to pass through the pore, and the other nucleic acid will be pushed away. The selectivity test in the presence of nucleic acids and different other microRNAs proved that this approach could be used in a complex solution (Tian et al. 2013).

(2) Solid-state nanopore

Besides biological nanopore, solid-state nanopores with all their diversity have also been commonly used to detect MicroRNA. Taking the example of micropipette nanopore, Hao Wang et al. used a micropipette nanopore to detect miRNA-21 as

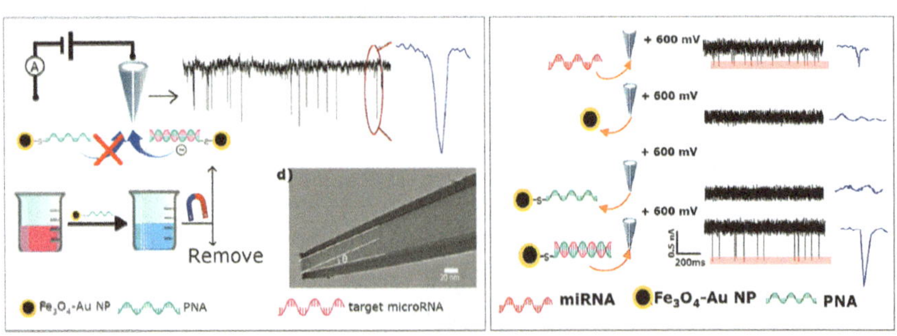


**Fig. 10** Working principle of the nanopipette nanopore sensor. Adapted with permission from Wang et al. (2019)

a lung cancer biomarker (Fig. 10). The team has used an innovative strategy, and the target microRNA was hybridized with neutral peptide nucleic acid (PNA) modified $Fe_3O_4$—Au nanoparticles, then the complex $Fe_3O_4$-Au-PNA-miRNA was translocated through the nanopipette nanopore by a positive potential application and eliminating neutral $Fe_3O_4$-Au-PNA. The authors were able to detect the miRNA-21 in the range of 2–50 nM in PBS with a limit of detection of 2 nM. Moreover, they demonstrated that their proposed sensor can be potentially applied to detect miRNA-21 in complicated samples (Wang et al. 2019). In the same context, Wanunu et al. have succeeded in detecting miR122a using a 5 nm pore drilled on a 10 nm thick silicon nitride membrane. In their platform, the miR122a was first hybridized with a specific probe, then the DNA probe/miRNA-122 was enriched by binding to the viral proteinp19-modified magnetic beads and finally translocated through the nanopore. In this work, they have studied several microRNA probes and the effect of the membrane thickness. Under the optimal conditions, the sensor could detect concentrations at the femtomolar level (Wanunu et al. 2010).

(3) Comparison between biological and solid-state nanopore

Although the sensing principle of the solid-state and biological nanopore is the same, the choice of an appropriate platform depending on the application and the experiment's chemical environment, the key differences between the two categories are the adjustment of the pore size, which is adjustable for the solid solid-state nanopores (synthetic nanopores) and not adjustable for the biological nanopores because the proteins' size is unchangeable, and they always keep the same size and shape. The other key difference is the resistivity of harsh chemical conditions; solid-state nanopores are more immune to rough chemical conditions. While biological nanopores are more suitable for surface modification (Shi et al. 2017). However, the first commercialized nanopore platform, and so far, the most used platform is based on biological nanopores delivered in 2014 by Oxford Nanopore Technology (Plesivkova et al. 2019).

## References

Angelucci, F., Cechova, K. Valis, M. Kuca, K. Zhang, B., & Hort, J. (2019). MicroRNAs in Alzheimer's disease: Diagnostic markers or therapeutic agents? *Frontiers in Pharmacology, 10*(JUN), 1–9.

Asangani, I. A., et al. (2008). MicroRNA-21 (miR-21) post-transcriptionally downregulates tumor suppressor Pdcd4 and stimulates invasion, intravasation and metastasis in colorectal cancer. *Oncogene, 27*(15), 2128–2136.

Bahri, M., Baraket, A., Zine, N., Ben Ali, M., Bausells, J., & Errachid, A. (2019). Capacitance electrochemical biosensor based on silicon nitride transducer for TNF-$\alpha$ cytokine detection in artificial human saliva: Heart failure (HF). Talanta.

Banzhaf-Strathmann, J., et al. (2014). Micro RNA -125b induces tau hyperphosphorylation and cognitive deficits in Alzheimer's disease. *The EMBO Journal, 33*(15), 1667–1680.

Bellassai, N., D'Agata, R., Jungbluth, V., & Spoto, G. (2019). Surface Plasmon Resonance for Biomarker Detection: Advances in Non-invasive Cancer Diagnosis. *Frontiers in Chemistry, 7*(August), 1–16.

Broughton, J. P., Lovci, M. T., Huang, J. L., Yeo, G. W., & Pasquinelli, A. E. (2016). Pairing beyond the seed supports MicroRNA targeting specificity. *Molecular Cell, 64*(2), 320–333.

Cai, B., Huang, L., Zhang, H., Sun, Z., Zhang, Z., & Zhang, G. J. (2015). Gold nanoparticles-decorated graphene field-effect transistor biosensor for femtomolar MicroRNA detection. *Biosensors and Bioelectronics, 74*, 329–334.

Calin, G. A., et al. (2002). Frequent deletions and down-regulation of micro-RNA genes miR15 and miR16 at 13q14 in chronic lymphocytic leukemia. *Proceedings of the National Academy of Sciences of the United States of America, 99*(24), 15524–15529.

Castañeda, A. D., Brenes, N. J., Kondajji, A. and Crooks, R. M. (2017). Detection of microRNA by electrocatalytic amplification: A general approach for single-particle biosensing. *Journal of the American Chemical Society, 139*(22), 7657–7664.

Castoldi, M., et al. (2006). A sensitive array for microRNA expression profiling (miChip) based on locked nucleic acids (LNA). *Rna, 12*(5), 913–920.

Chen, S. N., et al. (2019). MicroRNA in ovarian cancer: Biology, pathogenesis, and therapeutic opportunities. *International Journal of Environmental Research and Public Health, 16*(9), 1–14.

Cissell, K. A., Shrestha, S., & Deo, S. K. (2007). MicroRNA detection: Challenges for the analytical chemist. *Analytical Chemistry, 79*(13), 4754–4761.

Congur, G., Eksin, E. & Erdem, A. (2015). Impedimetric detection of microRNA at graphene oxide modified sensors. *Electrochimica Acta, 172*, 20–27.

Dave, V. P., et al. (2019). MicroRNA amplification and detection technologies: Opportunities and challenges for point of care diagnostics. *Laboratory Investigation, 99*(4), 452–469.

Erdem, A., Eksin, E., & Congur, G. (2015). Indicator-free electrochemical biosensor for microRNA detection based on carbon nanofibers modified screen printed electrodes. *Journal of Electroanalytical Chemistry, 755*, 167–173.

Erdem, A., Eksin, E., Isin, D., & Polat, D. (2017). Graphene oxide modified chemically activated graphite electrodes for detection of microRNA. *Electroanalysis, 29*(5), 1350–1358.

Feinbaum, R., Ambros, V., & Lee, R. (2004). The C. elegans Heterochronic Gene lin-4 Encodes Small RNAs with Antisense Complementarity to lin-14. *Cell, 75*(5), pp. 843–854.

Feng, J., Xing, W., & Xie, L. (2016). Regulatory roles of microRNAs in diabetes. *International Journal of Molecular Sciences, 17*(10), 1–12.

Gao, Z., Deng, H., Shen, W., & Ren, Y. (2013). A label-free biosensor for electrochemical detection of femtomolar microRNAs. *Analytical Chemistry, 85*(3), 1624–1630.

Gao, A., Lu, N., Dai, P., Li, T., & Wang, Y. (2013). Label-free and ultrasensitive detection of microrna biomarkers in lung cancer cells based on silicon nanowire FET biosensors. In *2013 Transducers and Eurosensors XXVII: The 17th International Conference on Solid-State Sensors, Actuators and Microsystems* (pp. 2439–2442). TRANSDUCERS and EUROSENSORS.

Gao, J., et al. (2020). Ultrasensitive label-free MiRNA sensing based on a flexible graphene field-effect transistor without functionalization. *ACS Applied Electronic Materials, 2*(4), 1090–1098.

Goh, S. Y., Chao, Y. X., Dheen, S. T., Tan, E. K., & Tay, S. S. W. (2019). Role of microRNAs in parkinson's disease. *International Journal of Molecular Sciences, 20*(22), 1–23.

Goodman, R. P., et al. (2005). Rapid chiral assembly of rigid DNA building blocks for molecular nanofabrication. *American Association for the Advancement of Science, 310*(5754), 1661–1665.

Guo, R., et al. (2018). Ultrasensitive simultaneous detection of multiplex disease-related nucleic acids using double-enhanced surface-enhanced Raman scattering nanosensors. *ACS Applied Materials and Interfaces, 10*(30), 25770–25778.

He, K., Liao, R., Cai, C., Liang, C., Liu, C., & Chen, X. (2016). Y-shaped probe for convenient and label-free detection of microRNA-21 in vitro. *Analytical Biochemistry, 499*, 8–14.

Huang, J., et al. (2019). Colorimetric and fluorescent dual-mode detection of microRNA based on duplex-specific nuclease assisted gold nanoparticle amplification. *Analyst, 144*(16), 4917–4924.

Hwang, M. T., et al. (2020). Ultrasensitive detection of nucleic acids using deformed graphene channel field effect biosensors. *Nature Communications*, *11*(1).

Isin, D., Eksin, E., & Erdem, A. (2017). Graphene oxide modified single-use electrodes and their application for voltammetric miRNA analysis. *Materials Science and Engineering C*, *75*, 1242–1249.

Ivica, J., Williamson, P. T. F., & de Planque, M. R. R. (2017). Salt gradient modulation of microRNA translocation through a biological nanopore. *Analytical Chemistry*, *89*(17), 8822–8829.

Jie, G., Zhao, Y., Wang, X., & Ding, C. (2017). Multiplexed fluorescence detection of microRNAs based on novel distinguishable quantum dot signal probes by cycle amplification strategy. *Sensors and Actuators, B: Chemical*, *252*, 1026–1034.

Joshi, G. K., Deitz-Mcelyea, S., Johnson, M., Mali, S., Korc, M., & Sardar, R. (2014). Highly specific plasmonic biosensors for ultrasensitive MicroRNA detection in plasma from pancreatic cancer patients. *Nano Letters*, *14*(12), 6955–6963.

Kamal Masud, M., et al. (2017). Gold-loaded nanoporous superparamagnetic nanocubes for catalytic signal amplification in detecting miRNA. *Chemical Communications*, *53*(8), 8231–8234.

Kaplan, M., Kilic, T., Guler, G., Mandli, J., Amine, A., & Ozsoz, M. (2017). A novel method for sensitive microRNA detection: Electropolymerization based doping. *Biosensors and Bioelectronics*, *92*, 770–778.

Kilic, T., Erdem, A., Erac, Y., Seydibeyoglu, M. O., Okur, S., & Ozsoz, M. (2015). Electrochemical detection of a cancer biomarker mir-21 in cell lysates using graphene modified sensors. *Electroanalysis*, *27*(2), 317–326.

Kilic, T., et al. (2012). Electrochemical based detection of microRNA, mir21 in breast cancer cells. *Biosensors and Bioelectronics*, *38*(1), 195–201.

Kilic, T., Nur Topkaya, S., & Ozsoz, M. (2013). A new insight into electrochemical microRNA detection: A molecular caliper, p19 protein. *Biosensors and Bioelectronics*, *48*, 165–171.

Kim, S. W., et al. (2010). A sensitive non-radioactive northern blot method to detect small RNAs. *Nucleic Acids Research*, *38*(7), 1–7.

Labib, M., Khan, N., Ghobadloo, S. M., Cheng, J., Pezacki, J. P., & Berezovski, M. V. (2013). Three-mode electrochemical sensing of ultralow MicroRNA levels. *Journal of the American Chemical Society*, *135*(8), 3027–3038.

Lee, K., et al. (2018). Recent progress in solid-state nanopores. *1704680*, 1–28.

Lee, J. H., Kim, J. A., Jeong, S., & Rhee, W. J. (2016). Simultaneous and multiplexed detection of exosome microRNAs using molecular beacons. *Biosensors and Bioelectronics*, *86*, 202–210.

Leggio, L., et al. (2017). MicroRNAs in parkinson's disease: From pathogenesis to novel diagnostic and therapeutic approaches. *International Journal of Molecular Sciences*, *18*(12).

Li, M., et al. (2019). Graphene oxide-based fluorometric determination of microRNA-141 using rolling circle amplification and exonuclease III-aided recycling amplification. *Microchimica Acta*, *186*(8).

Li, Y., Tian, R., Zheng, X., & Huang, R. (2016). Amplified electrochemical detection of nucleic acid hybridization via selective preconcentration of unmodified gold nanoparticles. *Analytica Chimica Acta*, *934*, 59–65.

Li, Q., Wang, Q., Yang, X., Wang, K., Zhang, H., and Nie, W. (2017). High sensitivity surface plasmon resonance biosensor for detection of microRNA and small molecule based on graphene oxide-gold nanoparticles composites. *Talanta*, *174*(June), 521–526.

Liu, L., et al. (2014). Highly sensitive and label-free electrochemical detection of microRNAs based on triple signal amplification of multifunctional gold nanoparticles, enzymes and redox-cycling reaction. *Biosensors and Bioelectronics*, *53*, 399–405.

Lu, Z., et al. (2016). Amplified voltammetric detection of miRNA from serum samples of glioma patients via combination of conducting magnetic microbeads and ferrocene-capped gold nanoparticle/streptavidin conjugates. *Biosensors and Bioelectronics*, *86*, 502–507.

Madison, B. B., et al. (2015). Let-7 represses carcinogenesis and a stem cell phenotype in the intestine via regulation of Hmga2. *PLoS Genetics*, *11*(8), 1–21.

Majd, S. M., Salimi, A., & Ghasemi, F. (2018). An ultrasensitive detection of miRNA-155 in breast cancer via direct hybridization assay using two-dimensional molybdenum disulfide field-effect transistor biosensor. *Biosensors and Bioelectronics, 105*, 6–13.

Miao, P., et al. (2016). Nuclease assisted target recycling and spherical nucleic acids gold nanoparticles recruitment for ultrasensitive detection of microRNA. *Electrochimica Acta, 190*, 396–401.

Mittal, S., Thakur, S., Mantha, A. K., & Kaur, H. (2019). Bio-analytical applications of nicking endonucleases assisted signal-amplification strategies for detection of cancer biomarkers -DNA methyl transferase and microRNA. *Biosensors and Bioelectronics, 124–125*, 233–243.

Mousavi, M. Z., et al. (2015). Urinary micro-RNA biomarker detection using capped gold nanoslit SPR in a microfluidic chip. *Analyst, 140*(12), 4097–4104.

Nie, W., et al. (2017). High sensitivity surface plasmon resonance biosensor for detection of microRNA based on gold nanoparticles-decorated molybdenum sulfide. *Analytica Chimica Acta, 993*, 55–62.

Ouyang, T., Liu, Z., Han, Z., & Ge, Q. (2019). MicroRNA detection specificity: Recent advances and future perspective. *Analytical Chemistry, 91*(5), 3179–3186.

Pang, Y., Wang, C., Wang, J., Sun, Z., Xiao, R., & Wang, S. (2016). $Fe_3O_4$@Ag magnetic nanoparticles for microRNA capture and duplex-specific nuclease signal amplification based SERS detection in cancer cells. *Biosensors and Bioelectronics, 79*, 574–580.

Peng, Y., & Gao, Z. (2011). Amplified detection of microRNA based on ruthenium oxide nanoparticle-initiated deposition of an insulating film. *Analytical Chemistry, 83*(3), 820–827.

Peng, Y., Yi, G., & Gao, Z. (2010). A highly sensitive microRNA biosensor based on ruthenium oxide nanoparticle-initiated polymerization of aniline. *Chemical Communications, 46*(48), 9131–9133.

Plesivkova, D., Richards, R., & Harbison, S. (2018). A review of the potential of the MinION TM single-molecule sequencing system for forensic applications. (December), 1–12, 2019.

Pritchard, C. C., Cheng, H. H., & Tewari, M. (2015). MicroRNA profiling: Approaches and considerations.*13*(5), 358–369.

Qin, S., & Zhang, C. (2011). MicroRNAs in vascular disease. *Journal of Cardiovascular Pharmacology, 57*(1), 8–12.

Rafiee-Pour, H. A., Behpour, M., & Keshavarz, M. (2016). A novel label-free electrochemical miRNA biosensor using methylene blue as redox indicator: Application to breast cancer biomarker miRNA-21. *Biosensors and Bioelectronics, 77*, 202–207.

Ramkissoon, S. H., Mainwaring, L. A., Sloand, E. M., Young, N. S., & Kajigaya, S. (2006). Non-isotopic detection of microRNA using digoxigenin labeled RNA probes. *Molecular and Cellular Probes, 20*(1), 1–4.

Rezaei, B., & Irannejad, N. (2019). Electrochemical detection techniques in biosensor applications. In *Electrochemical Biosensors* (pp. 11–43). Amsterdam: Elsevier.

Sapsford, K. E., Berti, L., & Medintz, I. L. (2006). Materials for fluorescence resonance energy transfer analysis: Beyond traditional donor-acceptor combinations. *Angewandte Chemie - International Edition, 45*(28), 4562–4589.

Sarkar, D., Liu, W., Xie, X., Anselmo, A. C., Mitragotri, S., & Banerjee, K. (2014). $MOS_2$ field-effect transistor for next-generation label-free biosensors. *ACS Nano, 8*(4), 3992–4003.

Schetter, A. J., Okayama, H., & Harris, C. C. (2012). The role of microRNAs in colorectal cancer. *Cancer Journal, 18*(3), 244–252.

Shen, W., Yeo, K. H., & Gao, Z. (2015). A simple and highly sensitive fluorescence assay for microRNAs. *Analyst, 140*(6), 1932–1938.

Shi, W., Friedman, A. K., & Baker, L. A. (2017). Nanopore Sensing.

Shu Zhu, C., et al. (2019). Avenues toward microrna detection in vitro: A review of technical advances and challenges. *Computational and Structural Biotechnology Journal, 17*, 904–916.

Shuai, H. L., Huang, K. J., Chen, Y. X., Fang, L. X., & Jia, M. P. (2017). Au nanoparticles/hollow molybdenum disulfide microcubes based biosensor for microRNA-21 detection coupled with duplex-specific nuclease and enzyme signal amplification. *Biosensors and Bioelectronics, 89*(Pt 2), 989–997.

Shuai, H. L., Huang, K. J., Xing, L. L., & Chen, Y. X. (2016). Ultrasensitive electrochemical sensing platform for microRNA based on tungsten oxide-graphene composites coupling with catalyzed hairpin assembly target recycling and enzyme signal amplification. *Biosensors and Bioelectronics*, *86*, 337–345.

Silahtaroglu, A., Pfundheller, H., Koshkin, A., Tommerup, N., & Kauppinen, S. (2004). LNA-modified oligonucleotides are highly efficient as FISH probes. *Cytogenetic and Genome Research*, *107*(1–2), 32–37.

Simões, F. R., & Xavier, M. G. (2017). Electrochemical sensors. In *Nanoscience and its Applications*, (pp. 155–178). Amsterdam: Elsevier Inc.

Singh, R., & Mo, Y. (2013). Role of microRNAs in breast cancer. *Cancer Biology & Therapy*, *14*(March), 201–212.

Song, R., et al. (2020). Detection of microRNA based on three-dimensional graphene field-effect transistor biosensor. *Nano*, *15*(3), 1–8.

Su, S., et al. (2016). On-electrode synthesis of shape-controlled hierarchical flower-like gold nanostructures for efficient interfacial DNA assembly and sensitive electrochemical sensing of microRNA. *Small*, *12*(28), 3794–3801.

Sultan, M., & Kanavarioti, A. (2019). Nanopore device-based fingerprinting of RNA oligos and microRNAs enhanced with an Osmium tag. *Scientific Reports* (September), 1–18.

Tan, L., Xu, L., Liu, J. W., Tang, L. J., Tang, H., & Yu, R. (2019). Duplex-specific nuclease-mediated target recycling amplification for fluorescence detection of microRNA. *Analytical Methods*, *11*(2), 200–204.

Tian, K., He, Z., Wang, Y., Chen, S.-J., & Gu, L. Q. (2013). Designing a polycationic probe for simultaneous enrichment and detection of microRNAs in a nanopore. *ACS Nano*, *7*(5), 3962–3969.

Tlili, C., et al. (2003). Fibroblast cells: A sensing bioelement for glucose detection by imedance spectroscopy. *Analytical Chemistry*, *75*(14), 3340–3344.

Tlili, C., Korri-Youssoufi, H., Ponsonnet, L., Martelet, C., & Jaffrezic-Renault, N. J. (2005). Electrochemical impedance probing of DNA hybridisation on oligonucleotide-functionalised polypyrrole. *Talanta*, *68*(1), 131–137.

Tran, H. V., Piro, B., Reisberg, S., Tran, L. D., Duc, H. T., & Pham M. C. (2013). Label-free and reagentless electrochemical detection of microRNAs using a conducting polymer nanostructured by carbon nanotubes: Application to prostate cancer biomarker miR-141. *Biosensors and Bioelectronics*, *49*, 64–169.

Tribolet, L., et al (2020). MicroRNA biomarkers for infectious diseases: From basic research to biosensing. *Frontiers in Microbiology*, *11*(June), 1–15.

Tyagi, S., & Kramer, F. R. (1996). Molecular beacons: Probes that fluoresce upon hybridization. *Nature Biotechnology*, *14*(3), 303–308.

Válóczi, A., Hornyik, C., Varga, N., Burgyán, J., Kauppinen, S., & Havelda, Z. (2004). Sensitive and specific detection of microRNAs by northern blot analysis using LNA-modified oligonucleotide probes. *Nucleic Acids Research*, *32*(22), e175.

Vilaivan, T. (2018). Fluorogenic PNA probes. *Beilstein Journal of Organic Chemistry*, *14*, 253–281.

Wang, H., Tang, H., Yang, C., & Li, Y. (2019). Selective single molecule nanopore sensing of microRNA using PNA functionalized magnetic core-shell $Fe_3O_4$-Au nanoparticles. *Analytical Chemistry*, *91*(12), 7965–7970.

Wanunu, M., Dadosh, T., Ray, V., Jin, J., McReynolds, L., & Drndić, M. (2010). Rapid electronic detection of probe-specific microRNAs using thin nanopore sensors. *Nature Nanotechnology*, *5*(11), 807–814.

Wanunu, M. (2012). Nanopores: A journey towards DNA sequencing. *Physics of Life Reviews*, *9*(2), 125–158.

Wen, S., et al. (2019). Plasmon coupling-enhanced raman sensing platform integrated with exonuclease-assisted target recycling amplification for ultrasensitive and selective detection of microRNA-21. *Analytical Chemistry*, *91*(19), 12298–12306.

Wightman, B., Ha, I., & Ruvkun, G. (1993). Posttranscriptional regulation of the heterochronic gene lin-14 by lin-4 mediates temporal pattern formation in C. elegans. *Cell*, *75*(5), 855–862.

Wu, K. L., Tsai, Y. M., Lien, C. T., Kuo, P. L., & Hung, J. Y. (2019). The roles of microRNA in lung cancer. *International Journal of Molecular Sciences*, *20*(7), 1–25.

Xi, Q., et al. (2014). Highly sensitive and selective strategy for microRNA detection based on $WS_2$ nanosheet mediated fluorescence quenching and duplex-specific nuclease signal amplification. *Analytical Chemistry*, *86*(3), 1361–1365.

Xu, F., et al. (2016). Ultrasensitive and multiple disease-related microRNA detection based on tetrahedral DNA nanostructures and duplex-specific nuclease-assisted signal amplification. *ACS Applied Materials and Interfaces*, *8*(49), 33499–33505.

Yang, H., et al. (2009). Direct, electronic microRNA detection for the rapid determination of differential expression profiles. *Angewandte Chemie - International Edition*, *48*(45), 8461–8464.

Yang, X., Wang, S., Wang, Y., He, Y., Chai, Y., & Yuan, R. (2018). Stimuli-responsive DNA microcapsules for SERS sensing of trace microRNA. *ACS Applied Materials and Interfaces*, *10*(15), 12491–12496.

Ye, J., Xu, M., Tian, X., Cai, S., & Zeng, S. (2019). Research advances in the detection of miRNA. *Journal of Pharmaceutical Analysis*, *9*(4), 217–226.

Yu, A.-M., & Pan, Y.-Z. (2012). Noncoding microRNAs: Small RNAs play a big role in regulation of ADME? *Acta Pharmaceutica Sinica B*, *2*(2), 93–101.

Zeng, Y., et al. (2017). Recent advances in surface plasmon resonance imaging: Detection speed, sensitivity, and portability. *Nanophotonics*, *6*(5), 1017–1030.

Zhang, J., et al. (2016). An immobilization-free electrochemical impedance biosensor based on duplex-specific nuclease assisted target recycling for amplified detection of microRNA. *Biosensors and Bioelectronics*, *75*, 452–457.

Zhang, H., Hiratani, M., Nagaoka, K., & Kawano, R. (2017). MicroRNA detection at femtomolar concentrations with isothermal amplification and a biological nanopore. *Nanoscale*, *9*(42), 16124–16127.

Zhang, H., Li, X., He, F., Zhao, M., & Ling, L. (2018). Turn-off colorimetric sensor for sequence-specific recognition of single-stranded DNA based upon Y-shaped DNA structure. *Scientific Reports*, *8*(1), 1–8.

Zhang, J., Li, Z., Wang, H., Wang, Y., Jia, H., & Yan, J. (2011). Ultrasensitive quantification of mature microRNAs by real-time PCR based on ligation of a ribonucleotide-modified DNA probe. *Chemical Communications*, *47*(33), 9465–9467.

Zhang, X., Wang, Y., Fricke, B. L., & Gu, L. Q. (2014). Programming nanopore ion flow for encoded multiplex microRNA detection. *ACS Nano*, *8*(4), 3444–3450.

Zhao, X., Zhao, Y., Deng, Y., Zhou, D., Zhang, Z., & Huang, Q. (2018a). DNA translocation through solid-state nanopore. no. March.

Zhao, G., et al. (2018b). Droplet digital PCR-based circulating microRNA detection serve as a promising diagnostic method for gastric cancer. *BMC Cancer*, *18*(1), 1–10.

Zhu, X., & Gao, T. (2018). *Spectrometry*. Amsterdam: Elsevier Inc.

# Early Detection of Helicobacter Pylori Bacteria in Complex Samples

**Hussamaldeen Jaradat, Ammar Al-Hamry, Mohammed Ibbini, and Olfa Kanoun**

**Abstract** Helicobacter Pylori (HP) bacteria is considered as one of the most capable pathogens in colonizing the human gastrointestinal tract. It is a dangerous carcinogen bacterium that infects about 50% of humans. Infection with HP may begin during childhood and can persist lifelong. Thus, detecting HP at an early stage is very important to prevent developing symptoms. Consequently, reliable detection techniques of HP for in-vitro samples are very important. In-vitro detection can include contaminated food, water, public sanitary facility, and any contaminated environment. Several techniques are available for detection of HP in complex samples, such as polymerase chain reaction (PCR) test and urea breath test (UBT). However, these techniques usually need excessive sample processing, complex instruments, trained personnel, and long detection time. Nowadays, detection of pathogens in complex samples is becoming very important as they are becoming more resistant to antibiotics. In this contribution we review detection methods of HP and address thereby important findings in stool antigen tests, fluorescent detection methods, calorimetric detection methods, Surface Plasmon Resonance detection methods as well as electrochemical methods. In addition, we provide perspectives for future developments in this important and challenging field.

H. Jaradat (✉) · A. Al-Hamry · O. Kanoun
Chair for Measurement and Sensor Technology, Technische Universität Chemnitz, Chemnitz, Germany
e-mail: hussam_bme@hotmail.com

A. Al-Hamry
e-mail: ammar.al-hamry@etit.tu-chemnitz.de

O. Kanoun
e-mail: olfa.kanoun@etit.tu-chemnitz.de

H. Jaradat
Department of Computer Engineering, Jordan University of Science and Technology, Irbid, Jordan

M. Ibbini
Department of Biomedical Engineering, Jordan University of Science and Technology, Irbid, Jordan
e-mail: mohib@just.edu.jo

 
O. Kanoun and N. Derbel (eds.), *Advanced Sensors for Biomedical Applications*, Smart Sensors, Measurement and Instrumentation 38,
https://doi.org/10.1007/978-3-030-71225-9_10

**Keywords** Helicobacter pylori · Bacteria detection · Sensors · Early detection · Nanomaterials

## 1 Introduction

Helicobacter pylori (HP) is a spirally-shaped gram-negative pathogen. It is a very dangerous bacteria such that infection with HP is strongly linked to gastrointestinal ulcers that may lead to cancer (Kusters et al. 2006). Mainly, it attacks stomach's epithelial cells targeting its apical surface, and with the help of its flagella tail-like which aids its movements, then it attaches itself through outer membrane proteins (Salama et al. 2013). Unlike other most bacteria, HP can persist lifelong in harsh acidic environment of human stomach (Königer et al. 2016). HP is very contagious, as it contaminated half of the world's population (Amieva et al. 2016). It worth noting that not all HP strains are severely pathogenic (Ali et al. 2019). The methods of transmission of HP may include human-to-human contact and contact to contaminated surfaces (Chauhan et al. 2018). Contaminated food, water and tools, untreated wastewater, toilets, and lack of hygienic habits are the main known pathways of HP transmission through indirect human interaction (Ng et al. 2017). Most HP infections occur during childhood and can begin to affect the host's life from that stage or later (Chen et al. 2019; Hong et al. 2019). Therefore, the possibility to detect HP at low concentrations, such as screening in contaminated mediums or in monitoring cases after eradication, is highly needed. In general, detection methods must be selective, sensitive, repeatable, reliable, cost-effective, and not time-consuming. Furthermore, compatibility with with point-of-care (POC) devices is a plus.

Several techniques available for HP detection both invasively and non-invasively such as: mass spectroscopy (MS), Giemsa stain, surface plasmon resonance (SPR), fluorescence techniques, polymerase chain reaction (PCR), and hematoxylin and eosin (H&E) stain, urea breath test (UBT), colorimetric and fluorescent-based biosensors, electrochemical biosensors, lateral flow devices (LFD) and enzyme-linked immunosorbent assay (ELISA) kits (Ali et al. 2019; Chen et al. 2019; Hong et al. 2019; Nosrati et al. 2017; Khalilpour et al. 2016). However, most of the proposed invasive and non-invasive techniques are used to diagnose human infections after symptoms have been developed (Nosrati et al. 2017). Besides, non-invasive detection techniques in complex samples generally suffer from several drawbacks such as the need for excessive sample preparation, long incubation period, poor sensitivity and/or specificity, the need for imaging and color comparison techniques and lack of complete quantitative results (Khalilpour et al. 2016).

For sensors to meet the quality criteria (ASSURED: Affordable, Sensitive, Specific, User-friendly, Rapid and Robust, Equipment-free and Deliverable to end-users). Kosack et al. (2017) formulated by the World Health Organization (WHO), this suggests careful choice of novel materials, novel biosensor structure and design, suitable fabrication techniques and efficient POC data analysis devices (Jaradat and Ibbini 2019). Therefore, there is still a current demand to produce simple but highly sensitive and specific HP sensors for complex clinical and environmental samples. Therefore, to achieve ultrasensitive biosensor enough for early detection, current strategies rely on using biomarkers and affinity sensing approach strategies without relying on PCR amplification step (Peng 2017). These approaches can include electrochemical, fluorescence, colorimetric, and SPR based biosensors.

This chapter highlights the following concerns about HP detection: firstly, the importance of in-Vitro detection of HP at early stage and for eradication monitoring; It highlights the necessity for developing a powerful ultrasensitive HP in-vitro detection method and illustrates its impact on fighting this pathogen. Secondly, the evaluation of recent in vitro detection methods: it criticizes the reliability of in-vitro detection methods proposed in literature, categorized according to the main transduction approach proposed by scientists and researchers, with each method's pros and cons are discussed. Thirdly, experience methodology and results of the authors are briefly discussed. In, addition, conclusions and future perspectives toward HP detection in complex sample are briefly summarized and discussed. Conclusion of the strengths and limitations for each detection approach are presented. Finally, it presents future perspectives and outlines for possible directions in HP early detection in complex sample.

## 2 Importance of HP In-Vitro Detection

This section illustrates the importance of detecting HP bacteria at very low concentrations in complex samples. Usually, the first exposition to HP can occur at a very early stage of human life, i.e. during childhood (Chen et al. 2019). In addition, HP re-infection monitoring after eradication starts with monitoring at very low bacterial culture count. On the other hand, contaminated mediums like food, water and communal facilities usually have very low concentration of HP (Khalilpour et al. 2016). With noting that scientists are reporting an increasing resistance of HP toward treatment with antibiotics, therefore, the detection of HP before infection or at early stage is very important in several ways (Khalilpour et al. 2016; Peng 2017). Firstly, early detection can help diagnose host children and help protect them from negative consequences of future possible infection with such dangerous pathogen. Especially that HP can cause irreversible damages to the GI tract and may lead to GI tract malfunctioning which may affect children's diet and nutrition throughout puberty (Toller et al. 2011). Secondly, the bacteria eradication using the traditional triple treatment is not a guarantee that the bacteria is not coming back because bacteria resistance for antibiotics is increasing (Mégraud 2012). Therefore, eradication success rate is dropping, thus, detecting the bacteria after the eradication will help in evaluating the eradication process itself, which is particularly important to determine the next medical intervention required (Mégraud 2012). Finally, the detection in complex samples at very low concentrations can help diagnose the contaminated mediums. This particularly helps in determining the routes of HP transmission. For example, simple HP in-vitro test can be incorporated in quality assurance routine tests such as testing supply water, vegetables or other food products for HP contamination. However, detecting HP contaminations of communal spaces may help to reduce the susceptibility of cross-infection between users of communal facilities. Thus, detecting HP in complex samples will greatly help in fighting disease spread and reducing its negative impact on human life.

# 3 Recent Advances in In-Vitro HP Detection

Since the discovery of HP in 1982, several detection techniques have been proposed to fight this pathogen (Ali et al. 2019). Most of these techniques were relying on using invasive methods and usually used to detect HP in patients with symptoms (Nosrati et al. 2017). Such invasive methods are beyond the scope of this chapter. Non-invasive in-vitro detection of HP is especially important for its casual simplicity, more comfortable for patients, more cost-effective, higher potential for use in POC devices and their need for less sample preparation (Jaradat and Ibbini 2019; Nosrati et al. 2017). UBT is very famous non-invasive detection method, in which a $C_{13}$ or $C_{14}$ labeled false urea should be drunk by the patient (Chey 2000). However, this method mainly relies on measuring the amount of $CO_2$ gas produced by bacteria activity. Although this test is non-invasive, it is not designed for early-stage detection nor for in-vitro detection. Because at low concentrations of bacteria, the production of $CO_2$ is very low and the test itself requires infrastructure.

## *3.1 HP Stool Antigen (HpSA) Test*

HP in-vitro detection methods usually depend on utilizing HP biomarkers like antigens, DNA and outer membrane proteins (OMP's) as evidence for existence of HP (Jaradat and Ibbini 2019). One of the famous in-vitro detection methods is the HP stool antigen (HpSA) test. This method depends on detecting the antigen which the human body produces as a response to HP infection with the help of PCR test as signal amplification aid (Shimoyama 2013). Stool, saliva, and blood serum can contain HpSA, and with the help of conventional serological techniques like ELISA, HSpA can be easily detected. Although its test is sufficiently simple and compatible with POC devices, recent studies started to question its reliability as a detection method because the number of false-negative results produced increases (Chen et al. 2019; Khalilpour et al. 2016). With utilizing PCR test for body antigen, HpSA can not be used to test contaminated medium like food and water. In addition, current detection technology trends are centered toward avoiding the PCR amplification.

### 3.2 Fluorescent Detection Methods

Fluorescent techniques are based on spectrochemical analysis method of fluorescence phenomena in which entities of an analyte will be excited by irradiation at a specific wavelength. Then the target entities will emit radiation at different wavelength, usually longer wavelength, which will be measured as a readout (Dolatabadi et al. 2011). Different fields of sciences widely use fluorescence techniques for several characteristics such as availability of several dyes and its inherent sensitivity. These properties put fluorescence as a promising sensing technique (Liu et al. 2017). Several dyes exist with four main families based on its synthesis: peptides and proteins, synthetic polymers, and multi-component compounds (Liu et al. 2013; Stockert and Blázquez-Castro 2017).

Fluorescent dyes based on synthetic semiconductor quantum dots (QD's) can be utilized as motifs for antibodies to function as sensing probes (Stockert and Blázquez-Castro 2017). QD's are characterized by SPR, because they emit a specific wavelength and angle of radiation when exposed to UV-Vis electromagnetic waves. QD's are commonly functionalized with biological receptors like DNA strands and antibody antigens for biosensor applications (Feng et al. 2014). Therefore, attaching the QD's to the receptor antibody-antigen will change the surface properties of the QD's affecting their fluorescence properties (Wang et al. 2010). Thus, when HP antibody immunogenically interacts with the recipient antigen, another surface change occurs modulating the radiation angle, the intensity of radiation, or wavelength of the fluorescent wave (Feng et al. 2014; Nezami et al. 2017; Wang et al. 2010). Quantifying such fluctuations can lead to specifically quantifying the bacteria's existence and concentration.

Recent literature reports several fluorescent sensors and proof of concepts for in-vitro detection of HP. For example, Liu et al. (2017) reported a fluorescent sensor based on CuInS2 QD conjugated with single-stranded HP DNA (ssDNA) aptamer bounded to graphene oxide (GO) substrate. In this sensor, if HP complementary DNA is absent, then the ssDNA is adsorbed on GO which quenches the QD fluorescence (Liu et al. 2017). But if the complementary HP DNA is introduced then ssDNA will bound to it to form a double-stranded DNA (dsDNA). Formation of dsDNA disturbs the adsorption resulting in restoring the fluorescence activity of QD but at different intensity.

A novel method proposed by Chen et al. (2019) by using immunomagnetic beads (IMP's) bonded to monoclonal antibody (mAbs). mAbs attached with IMP were utilized to immunogenically capture and magnetically extract HP from the complex sample. The immunocomplex then eluted and exposed to fluorescence QD's conjugated with polyclonal antibodies (pAbs) probe. The interaction of pAbs/QD with HP alters the fluorescence of HP. Hong et al. (2019) proved the possibility to use a pH-sensitive fluorescence probe based on benzothiazole conjugated with hydroxythiophene ($T_2$(OH)B). This sensor relies on the fact that HP secrets urease enzyme which is able to neutralize the acidic surrounding of the bacteria by catalyzing urea hydrolysis to enable its survival in the harsh stomach environment. As a result, PH is locally altered resulting in altering the fluoresce of the fluorescent probe.

Akashi et al. (2019) proposed a detection method based on $\gamma$-glutamyl hydroxymethyl rhodamine green fluorescent probe (gGlu-HMRG). This probe can react with HP $\gamma$-glutamyl transpeptidase (GGT), produced by HP, to produce immediate fluorescence. Therefore, gGlu-HMRG proved to detect infection of HP in ex-vivo gastric biopsy. The activity of HP to produce GGT was quantified by this fluorescent probe assay by determining fluorescence intensity changes. In this work, gGlu-HMRG in HP-negative cultures resulted in considerably less fluorescence than in HP-positive samples after incubation for 15 min. Although the assay sensitivity and specificity were lower than 90%, this assay can be promising in rapid HP detection for ex-vivo specimen.

HP detection in complex samples using fluorescence-based strategies still require bulky instruments, or at the most portable end, still needs imaging technologies such as smartphone equipped with a proper toolkit and software, especially when detection is at low concentrations or quantification is required (Chen et al. 2019; Dolatabadi et al. 2011; Jaradat and Ibbini 2019; Nosrati et al. 2017; Liu et al. 2013). Therefore, fluorescent techniques generally have lower chances to be effectively utilized as precise quantitative detection methods especially in remote regions. In addition, it usually needs relatively longer test durations (Akashi et al. 2019; Hong et al. 2019; Liu et al. 2017).

### *3.3 Colorimetric Detection Methods*

Colorimetric analysis methods determine the existence and concentration of analyte using a color reagent (Zarei et al. 2017). Its application ranges from inorganic to organic and biological samples (Vilela et al. 2012). Biological analyte detection using colorimetric sensor usually relies on a shift of absorbed or reflected light intensity when a biologically labeled-reagent complex interacts with the analyte (Vilela et al. 2012). Colorimetric properties alternation is due to change of physical properties of the probe upon interaction with the target (Ali et al. 2019).

Few colorimetric sensors for detection of HP in complex sample without PCR amplification were reported by researchers. For example, Ali et al. (2019) reported a colorimetric-based sensing method utilizing a protein-activated DNA/RNA-cleaving property of fluorescent DNAzyme (deoxyribozymes or DNA enzymes) activated by HP protein. Where DNAzymes are synthetic single-stranded DNA (ssDNA) molecules able to catalyze specific chemical reaction. DNA-RNA substrate cleavage occurs at a specific ribonucleotide junction (R) in a close proximity of a quencher (Q) and a fluorophore (F) where it suppresses the fluorescence to its minimum, with the existence of a quencher (Tram et al. 2012). But once the protein assisted cleavage occurs, the quencher is released resulting in much stronger fluorescence. In Ali et al. (2019), DHp3T4 is the DNA segment with full cleavage activity but with shortest DNAzyme sequence possible. HP protein is obtained using the crude extracellular mixture (CEM) methodology, in which residue entities left behind by the bacteria in its complex sample are collected and used as HP specific activator. In their work, DHp3T4 is bound with urease and attached to agarose beads for immobilization and after addition of HP CEM to the sensing assay, it can result in freeing urease which is collected by centrifugation. Adding the extract to phenol-red and urea solution results in urease hydrolyzing urea which releases ammonia that makes the medium to become more basic (higher pH) which changes the fluorescent color from yellow to red . This methodology uses stool sample with very good limit of detection (LOD) of $10^4$cf umL$^{-1}$ (Ali et al. 2019). They also prepared a paper-based version of the sensing platform with semi-quantitative results.

## *3.4 Surface Plasmon Resonance Detection Methods*

Biosensors based on SPR are extensively utilized as a powerful method to monitor biological interactions in real-time, mainly, due to the highly specific binding with biological entities (Wang et al. 2016). Label-free and high-throughput are the main intrinsic properties of SPR technique in measuring affinity interactions and/its kinetics (Wang et al. 2016). In SPR sensors, biological entity attached to a nanostructure surface will serve as receptor for the analyte. The interaction of the receptor and analyte will locally alter the refractive index in the interaction vicinity on the nanostructured surface (Lazcka et al. 2007). The shift in refractive index value is measured by SPR sensor apparatus (Nishimura et al. 2000).

Nishimura et al. (2000), made use of immunological response between HP urease and its monoclonal antibodies utilizing a commercially available phenol-red based assay kit. The kit utilizes the binding behavior of HP's urease monoclonal antibodies attached on 50-nm thick gold-coated sensor chip. Due to bacteria surface comprised of several proteins and biological structures, the SPR-based apparatus utilized a measuring cell for the sample containing HP and another reference cell for a sample without HP cells to neutralize the non-specific binding effect. In this study, it was reported that the LOD reached down to $2 \times 10^7$ bacteria/ml.

In another study, Fabini et al. (2016) proposed an isothermal titration calorimetry system with SPR capabilities based on monitoring the binding of HP's nickel-dependent regulatory protein (NikR) (HpNikR) through Ni(II) with DNA. However, this measurement method requires complicated setup and not very suitable for POC applications and more oriented to drug research (Fabini et al. 2016). SPR methodologies have significant limitations such as cross-sensitivity to molecules with analyte-similarly-structured molecules, non-specific SPR response on sensor surface from non-target biological entities, intrinsic alteration in refractive index due to environmental and compositional oscillation of the complex samples, and non-suitability for POC as it requires complex devices and setup (Nosrati et al. 2017; Wang et al. 2016; Lazcka et al. 2007; Nishimura et al. 2000; Fabini et al. 2016).

## 3.5 Electrochemical Detection Methods

Electrochemical transduction based nano-biosensors usually rely on surface engineering of an electrochemical probe to achieve sensitive and specific bio-sensitivity (Jaradat and Ibbini 2019). Electrochemical biosensors exhibit amazing performance with respect to other competitive sensing approaches (Talebi Bezmin Abadi 2014). It possesses inherently superior sensitivity, better reproducibility, and more portability (Kaushik et al. 2018; Smith et al. 2019). Standard electrochemical-transduction-based sensor uses the three-electrode configuration: working electrode (WE), reference electrode (RE) and counter electrode (CE) (Makableh et al. 2020). In these electrodes, the WE and CE are surface engineered and can be functionalized using the same nanotechnological approach. However, further functionalization could be applied on WE surface by utilizing bio-affinity receptors to enhance specificity toward biomarker analytes (Maduraiveeran et al. 2018; Makableh et al. 2020). Cyclic voltammetry (CV), differential pulse voltammetry (DPV), square wave voltammetry (SWV) and electrochemical impedance spectroscopy (EIS) are typical electrochemical measurement setups used in sensing applications . Electrochemical detection of HP in complex samples usually utilizes ssDNA hybridization or protein thorough antigen-antibody reactions (Maduraiveeran et al. 2018).

Recently, several electrochemical biosensors have been proposed, for e.g. Peng (2017) introduced an electrochemical gold electrode functionalized with single-stranded DNA probe (ssDNA) biosensor, with $\beta$-cyclodextrin ($\beta$-CD) DNA as electroactive material. In this senor, occurrence of DNA hybridization results in accumulating the $\beta$-CD on sensor surface. The proposed sensor detected HP in excrement using DPV measurement technique with results comparable to ELISA strategy at concentrations as low as 5 nM.

In another recent study, Jain et al. (2020) produced HP biosensor utilizing nanomaterials and cytotoxin-associated gene (CagA) antibody-antigen. This electrochemical sensor is label-free immunosensor based on triple-structured sensor surface designed for the noninvasive HP detection in stool sample (Jain et al. 2020). They utilized an Au-bare electrode electropolymerized with the biologically compatible polyindole carboxylic acid (Pin5COOH) which is recognized for its good electroconductivity. Followed by deposition of carboxylated multi-walled carbon nanotubes (c-MWCNT) layer where the c-MWCNT solution was improved by ultrasonication (Jain et al. 2020). c-MWCNTs is a well-known semiconductor material for its superior galvanic and mechanical properties with large surface area enhancing electronic transfer at electrode surface (Makableh et al. 2020). The carboxylated terminal of MWCNTs is utilized in immobilizing antibodies by crosslinking (Maduraiveeran et al. 2018). It was followed by Electrodeposition of titanium oxide nanoparticles ($TiO_2$) nanoparticles to utilize its enhanced biological conjugation and stability at long-term (Jain et al. 2020). For specificity, the authors further cross-linked the c-MWCNTs on sensor surface with anti-CagA antibody. The electrochemical characterization was performed using EIS, SWV and CV, which resulted in sensor LOD as low as 0.1 ng/mL with sensor half-life of 6 weeks.

Electrochemical sensing utilizes potentiostat devices for analysis resulting in being more suitable for POC applications (Fort et al. 2019). In addition, electrochemical sensors usually use minimal sample and/or reagent amounts without the need for bulky instruments resulting in more affordable cost (Essousi et al. 2019). Moreover, electrochemical sensors fabrication is more cost-effective by utilizing the screen-printed electrode technology enabling fully-functional disposable electrodes. Electrochemical sensors are inherently ultrasensitive and when combined with nanotechnology and nanomaterials may result in enabling early detection of HP in complex clinical samples and without the need for PCR preamplification step.

## 4 Conclusions and Future Perspective

HP is a very serious and dangerous pathogen that can cause several gastrointestinal disorders that may lead to cancer. Detecting HP at early stages, after eradication and in contaminated in-vitro environments is very important to control and fight this persistent pathogen. Several biosensors with different approaches and methodologies have been proposed for HP detection in complex samples based on optical, electrochemical and ELISA based methods. Optical methods like, fluorescence, SPR and colorimetric based sensing, have a high sensitivity and excellent selectivity. However, such approaches usually should be backed-up with bulky optical analysis devices. Additionally, they need a long test time. Apart from that, electrochemical-based biosensors have inherently higher sensitivity that enables detection at very low concentrations which is useful for early detection. Their cost-effective property is making it a good candidate for POC devices, especially for their simple handling, except that they still suffer from false-positive results due to complex sample electrolytes. Table 1 summarizes different sensing techniques and their performance.

Consequently, HP biosensor that is compatible with the ASSURED standard for healthcare outlined by WHO in complex samples still a major challenge. Thus, recent research trends focus on developing biosensor that are sensitive, selective, reproducible, rapid, more affordable, compatible with POC units and able to work in complex sample like stool, saliva food and water. Possible routes to enhance HP biosensors range from using more unique HP biomarkers to enhance sensor specificity with less sample pretreatment. Successful HP biosensor can greatly help fighting HP spread and may be included in antibiotic resistance studies of pathogens in general. Successful pathogen biosensors can be utilized in quality assurance routine tests in several facilities such as factories, hospitals, restaurants, water management stations as well as personal care.

**Table 1** Limit of detection LOD, incubation time and linearity range LR of detection methods for of HP bacteria

| Detection method | LOD | Linear range | Detection time | Reference |
|---|---|---|---|---|
| Optical-fluorescent | 0.46 pM | 1.25–875 pM | ~80 min | Liu et al. (2017) |
| Optical-fluorescent | $10^2$ CFU* $mL^{-1}$ | $10^{-1} - 10^6$ CFU $mL^{-1}$ | ~120 min | Chen et al. (2019) |
| Optical-fluorescent | 190 nM | Not reported | ~60 min | Hong et al. (2019) |
| Optical-fluorescent | Not reported | Not reported | ~15 min | Akashi et al. (2019) |
| Optical-colorimetric | $10^4$ CFU $mL^{-1}$ | $10^3 - 10^6$ CFU $mL^{-1}$ | ~20 min | Ali et al. (2019) |
| Optical-SPR | $2 \times 10^6$ bacteria/ml | Not reported | Not reported | Nishimura et al. (2000) |
| Electro-chemical | 0.15 nM | 0.3 nM–0.24 μM | >90 min | Peng (2017) |
| Electro-chemical | ~0.77 pM | 0.77–61.5 pM | >12 min | Jain et al. (2020) |

*Colony-forming unit (i.e. only viable bacteria cells are counted)

## References


Amieva, M., & Peek, R. M, Jr. (2016). Pathobiology of helicobacter pylori-induced gastric cancer. *Gastroenterology, 150*(1), 64–78.

Akashi, T., Isomoto, H., Matsushima, K., Kamiya, M., Kanda, T., & Nakano, M., et al. (2019). A novel method for rapid detection of a Helicobacter pylori infection using a γ-glutamyltranspeptidase-activatable fluorescent probe. *Scientific Reports, 9*(1).

Ali, M. M., Wolfe, M., Tram, K., Gu, J., Filipe, C. D. M., Li, Y., et al. (2019). A DNAzyme-based colorimetric paper sensor for helicobacter pylori. *Angewandte Chemie International Edition, 58*(29), 9907–9911.

Chauhan, N., Gupta, S., Avasthi, D. K., Adelung, R., Mishra, Y. K., & Jain, U. (2018). Zinc oxide tetrapods based biohybrid interface for voltammetric sensing of helicobacter pylori. *ACS Applied Materials and Interfaces, 10*(36), 30631–30639.

Chen, L., Li, X., Zou, T., Wang, T., Cui, X., Chen, Y., Zhang, C., & Zhao, S. (2019). Ultrasensitive detection of H. pylori in human feces based on immunomagnetic bead capture and fluorescent quantum dots. *The Analyst, 144*(13), 4086–4092.

Chey, W. D. (2000). Accurate diagnosis of helicobacter pylori. *Gastroenterology Clinics of North America, 29*(4), 895–902.

Dolatabadi, J. E. N., Mashinchian, O., Ayoubi, B., Jamali, A. A., Mobed, A., Losic, D., et al. (2011). Optical and electrochemical DNA nanobiosensors. *TrAC Trends in Analytical Chemistry, 30*(3), 459–472.

Essousi, H., Barhoumi, H., Bibani, M., Ktari, N., Wendler, F., Al-Hamry, A., et al. (2019). Ion-imprinted electrochemical sensor based on copper nanoparticles-polyaniline matrix for nitrate detection. *Journal of Sensors, 2019*, 1–14.

Fabini, E., Zambelli, B., Mazzei, L., Ciurli, S., & Bertucci, C. (2016). Surface plasmon resonance and isothermal titration calorimetry to monitor the Ni(II)-dependent binding of Helicobacter pylori NikR to DNA. *Analytical and Bioanalytical Chemistry, 408*(28), 7971–7980.

Feng, C., Dai, S., & Wang, L. (2014). Optical aptasensors for quantitative detection of small biomolecules: A review. *Biosensors and Bioelectronics*, *59*, 64–74.

Fort, A., Panzardi, E., Al-Hamry, A., Vignoli, V., Mugnaini, M., Addabbo, T., et al. (2019). Highly sensitive detection of NO2 by Au and TiO2 nanoparticles decorated SWCNTs sensors. *Sensors*, *20*(1), 12.

Hong, K.-I., Park, S.-H., Lee, S. M., Shin, I., & Jang, W.-D. (2019). A pH-sensitive excited state intramolecular proton transfer fluorescent probe for imaging mitochondria and Helicobacter pylori. *Sensors and Actuators B: Chemical*, *286*, 148–153.

Jain, U., Gupta, S., Soni, S., Khurana, M. P., & Chauhan, N. (2020). Triple-nanostructuring-based noninvasive electro-immune sensing of CagA toxin for Helicobacter pylori detection. *Helicobacter*, *25*(4).

Jaradat, H., & Ibbini, M. (2019). Helicobacter pylori detection methods in complex samples: A mini-review. *International Journal of Biosensors and Bioelectronics*, *5*(5).

Kaushik, A., Yndart, A., Kumar, S., Jayant, R. D., Vashist, A., & Brown, A. N., et al. (2018). A sensitive electrochemical immunosensor for label-free detection of Zika-virus protein. *Scientific Reports*, *8*(1).

Khalilpour, A., Kazemzadeh-Narbat, M., Tamayol, A., Oklu, R., & Khademhosseini, A. (2016). Biomarkers and diagnostic tools for detection of helicobacter pylori. *Applied Microbiology and Biotechnology*, *100*(11), 4723–4734.

Königer, V., Holsten, L., Harrison, U., Busch, B., Loell, E., & Zhao, Q., et al. (2016). Helicobacter pylori exploits human CEACAMs via HopQ for adherence and translocation of CagA. *Nature Microbiology*, *2*(1).

Kosack, C. S., Page, A.-L., & Klatser, P. R. (2017). A guide to aid the selection of diagnostic tests. *Bulletin of the World Health Organization*, *95*(9), 639–645.

Kusters, J. G., van Vliet, A. H. M., & Kuipers, E. J. (2006). Pathogenesis of helicobacter pylori infection. *Clinical Microbiology Reviews*, *19*(3), 449–490.

Ng, C. G., Loke, M. F., Goh, K. L., Vadivelu, J., & Ho, B. (2017). Biofilm formation enhances helicobacter pylori survivability in vegetables. *Food Microbiology*, *62*, 68–76.

Nosrati, R., Golichenari, B., Nezami, A., Taghdisi, S. M., Karimi, B., Ramezani, M., et al. (2017). Helicobacter pylori point-of-care diagnosis: Nano-scale biosensors and microfluidic systems. *TrAC Trends in Analytical Chemistry*, *97*, 428–444.

Liu, Z., & Su, X. (2017). A novel fluorescent DNA sensor for ultrasensitive detection of helicobacter pylori. *Biosensors and Bioelectronics*, *87*, 66–72.

Lazcka, O., Campo, F. J. D., & Muñoz, F. X. (2007). Pathogen detection: A perspective of traditional methods and biosensors. *Biosensors and Bioelectronics*, *22*(7), 1205–1217.

Liu, J., Liu, C., & He, W. (2013). Fluorophores and their applications as molecular probes in living cells. *Current Organic Chemistry*, *17*(6), 564–579.

Maduraiveeran, G., Sasidharan, M., & Ganesan, V. (2018). Electrochemical sensor and biosensor platforms based on advanced nanomaterials for biological and biomedical applications. *Biosensors and Bioelectronics*, *103*, 113–129.

Makableh, Y. F., Al-Fandi, M., Jaradat, H., Al-Shami, A., Rawashdeh, I., & Harahsha, T. (2020). Electrochemical characterization of nanosurface-modified screen-printed electrodes by using a source measure unit. *Bulletin of Materials Science*, *43*(1).

Mégraud, F. (2012). The challenge of Helicobacter pylori resistance to antibiotics: The comeback of bismuth-based quadruple therapy. *Therapeutic Advances in Gastroenterology*, *5*(2), 103–109.

Nezami, A., Nosrati, R., Golichenari, B., Rezaee, R., Chatzidakis, G. I., Tsatsakis, A. M., et al. (2017). Nanomaterial-based aptasensors and bioaffinity sensors for quantitative detection of 17$\beta$-estradiol. *TrAC Trends in Analytical Chemistry*, *94*, 95–105.

Nishimura, T., Hifumi, E., Fujii, T., Niimi, Y., Egashira, N., Shimizu, K., et al. (2000). Measurement of helicobacter pylori using anti its urease monoclonal antibody by surface plasmon resonance. *Electrochemistry*, *68*(11), 916–919.

Peng, P. (2017). Fabrication of an electrochemical sensor for helicobacter pylori in excrement based on a gold electrode. *International Journal of Electrochemical Science*, 9478–9487.

Salama, N. R., Hartung, M. L., & Müller, A. (2013). Life in the human stomach: Persistence strategies of the bacterial pathogen helicobacter pylori. *Nature Reviews Microbiology*, *11*(6), 385–399.

Shimoyama, T. (2013). Stool antigen tests for the management of Helicobacter pylori infection. *World Journal of Gastroenterology*, *19*(45), 8188. https://doi.org/10.3748/wjg.v19.i45.8188.

Smith, S., Madzivhandila, P., Ntuli, L., Bezuidenhout, P., Zheng, H., & Land, K. (2019). Printed paper-based electrochemical sensors for low-cost point-of-need applications. *Electrocatalysis*, *10*(4), 342–351.

Stockert, J.C., & Blázquez-Castro, A. (2017). *Chapter 4: Fluorescent labels. Fluorescence microscopy in life sciences* (pp. 96–134). Bentham Science Publishers. ISBN 978-1-68108-519-7. Retrieved 24 December 2017.

Talebi Bezmin Abadi, A. (2014). Helicobacter pylori: Emergence of a superbug. *Frontiers in Medicine, 1*.

Toller, I. M., Neelsen, K. J., Steger, M., Hartung, M. L., Hottiger, M. O., Stucki, M., et al. (2011). Carcinogenic bacterial pathogen helicobacter pylori triggers DNA double-strand breaks and a DNA damage response in its host cells. *Proceedings of the National Academy of Sciences*, *108*(36), 14944–14949.

Tram, K., Kanda, P., & Li, Y. (2012). Lighting Up RNA-Cleaving DNAzymes for Biosensing. *Journal of Nucleic Acids*, *2012*, 1–8.

Vilela, D., González, M. C., & Escarpa, A. (2012). Sensing colorimetric approaches based on gold and silver nanoparticles aggregation: Chemical creativity behind the assay. *A Review. Analytica Chimica Acta*, *751*, 24–43.

Wang, G., Wang, Y., Chen, L., & Choo, J. (2010). Nanomaterial-assisted aptamers for optical sensing. *Biosensors and Bioelectronics*, *25*(8), 1859–1868.

Wang, H., Wang, X., Wang, J., Fu, W., & Yao, C. (2016). A SPR biosensor based on signal amplification using antibody-QD conjugates for quantitative determination of multiple tumor markers. *Scientific Reports*, *6*(1).

Zarei, M. (2017). Portable biosensing devices for point-of-care diagnostics: Recent developments and applications. *TrAC Trends in Analytical Chemistry*, *91*, 26–41.

The manufacturer's authorised representative in the EU is Springer Nature Customer Service Centre GmbH, Europaplatz 3, 69115 Heidelberg, Germany. If you have any concerns regarding our products, please contact ProductSafety@springernature.com

Printed and bound by CPI Group (UK) Ltd, Croydon, CR0 4YY
30/09/2026
02225306-0018